The [illegible] Management Blueprint

How Any Beginner Can Thrive as a Successful Project Manager with This Stress-Free, Step-by-Step Guide to Mastering the Essentials

Franklin Publications

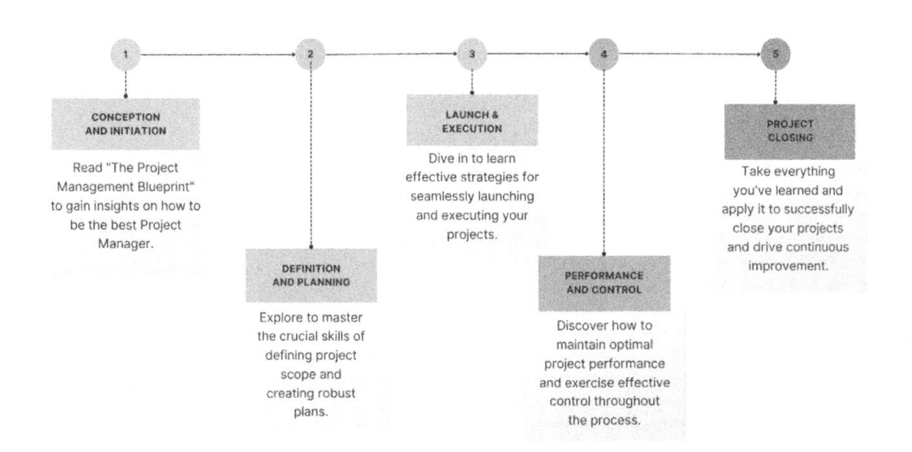

Contents

Introduction

My fascination with project management started as a blend of curiosity and necessity. At the beginning of my career, I unexpectedly found myself forced into the role of Project Manager despite missing formal certifications, substantial training, or, quite frankly, the corresponding salary that goes alongside it. Since then, I've led numerous projects worldwide, with budgets reaching billions. I quickly learned that while the responsibility is great, the rewards of effectively steering a project from idea to reality are even more significant. Driven by a passion for demystifying the complex world of project management, I embarked on a mission to make this field accessible and achievable for everyone. Whether you are not a Project Manager, a fresh "recruit," or a seasoned professional seeking to polish your skills, this book is crafted for you.

'The Project Management Blueprint' is not just another manual; it is a practical roadmap designed to transform the intimidating task of becoming a project manager into a transparent and manageable process. Here, you will find a unique combination of practical advice, real-world examples, and innovative strategies tailored to guide beginners and seasoned experts. This guide is not just about theory; it's about equipping you with the practical skills and knowledge you need to succeed in project management.

Project management is not just a career; it's a gateway to personal growth and professional advancement. It's one of the fastest-growing fields, offering lucrative opportunities across technology, healthcare, finance, and construction industries. With organizations increasingly relying on skilled project managers to navigate complex projects and drive success, the demand for professionals with the right skills is skyrocketing. This book positions you to seize these opportunities, providing a gateway to a rewarding career filled with growth and advancement.

Our primary readers are individuals like you—beginners eager to step into the role of project manager, working adults pursuing a significant career shift, and experienced managers aiming to stay abreast of the latest methodologies and technological advances. This book is not just a guide; it's a companion on your journey. It assures you that regardless of your starting point, whether you're a complete novice or a seasoned professional, you will be equipped to navigate the evolving landscape of project management. Your journey is important to us, and we're here to support you every step of the way.

Success in this field requires more than just a grasp of tools and techniques; it demands adaptability, resilience, and a commitment to learning. These themes are woven throughout the book, preparing you to tackle challenges and capitalize on opportunities with confidence and expertise.

Structured to guide you from foundational concepts to advanced strategies, the book also focuses on special topics like the future of project management and the transformative role of technology and innovation. This journey equips you to participate in and lead some of the world's most exciting projects.

I invite you to join me on this learning adventure. With each chapter, you'll gain knowledge and skills and the confidence to apply them to your projects. By the end of this book, envision yourself not just as a project manager but as a visionary capable of turning the most daunting challenges into outstanding successes. At the end of the book, I've also added the "Top 100 Definitions Used in Project Management". So, take your time to learn some of the vocabulary in the book. These definitions will serve as a handy reference guide, helping you navigate the terminology and concepts we'll explore throughout our journey together.

Throughout this book, we'll also explore the essential soft skills that often set exceptional project managers apart. While technical know-how is crucial, the ability to communicate clearly, lead diverse teams, resolve conflicts, and make tough calls under pressure can truly make or break a project. We'll work on developing these skills together, using real-world examples and practical exercises to sharpen your leadership and emotional intelligence. Given today's global business landscape, we'll also touch on the importance of cultural awareness, preparing you to navigate projects across different cultures and time zones. My goal is to help you become a well-rounded professional who can handle any project thrown your way. After all, project management is just as much about working with people as it is about managing tasks, and I want to make sure you're ready for both challenges.

Let's begin this transformative journey together, stepping confidently towards a future where you are the architect of innovation and success in the captivating world of project management.

Chapter 1

Laying the Foundation

Every great project starts with a solid foundation. In project management, this foundation is built on a clear understanding of what project management entails and its pivotal roles and responsibilities. Whether it's a groundbreaking architectural marvel or a crucial software update, project management is the backbone that supports and guides a project from an idea to reality. In this chapter, we will explore the core aspects of project management, providing you with the essential knowledge and skills needed to begin your path toward becoming a proficient project manager.

1.1 What is Project Management? Understanding the Role and Responsibilities

At its core, project management is the discipline of organizing, planning, securing, managing, leading, and controlling resources to achieve specific goals. A project is a temporary endeavor with a defined beginning, end, scope, and resources. Due to their finite nature, projects are distinct from routine operations, necessitating distinct management techniques and strategies.

The project management scope spans various industries, from constructing skyscrapers and managing healthcare initiatives to developing new technologies and rolling out software updates. For instance, consider the construction of a bridge, which requires meticulous planning to manage budgets, timelines, materials, and human resources. Similarly, in the IT sector, rolling out new software or updates involves coordinating numerous tasks to ensure that the final product meets customer expectations and remains within budget and schedule constraints.

The Role of a Project Manager

A project manager's role is multifaceted and dynamic. At the heart of this role is ensuring that a project is completed on time, within budget, and to the required quality standards. This involves a wide range of activities, from the initial planning stages to the final touches that bring a project to completion. Critical aspects of a project manager's role include:

- **Planning** involves setting goals, defining roles, and producing schedules and timelines for tasks in consultation with other team members.

- **Executing**: Involves managing teams, allocating resources effectively, and ensuring that tasks are carried out according to plan.

- **Monitoring**: This includes tracking a project's progress, verifying that it remains on track, and keeping stakeholders informed.

- **Controlling**: Involves making adjustments to the project plan to accommodate changes or shifts in the project's scope.

- **Closing**: The project manager ensures that the project is completed correctly, including settling accounts, resolving any pending details, and analyzing the project's success.

Project managers, with their robust leadership, communication, negotiation, problem-solving, and risk-management skills, have the power to transform projects and organizations. These competencies are not just crucial for project success, but also for personal growth. For instance, leadership skills empower you to guide teams and make strategic decisions, while communication skills enable you to effectively convey project updates and manage stakeholder expectations. Problem-solving and risk-management skills give you the ability to identify and address project issues and mitigate potential risks, thereby transforming challenges into opportunities.

Responsibilities Breakdown

Effective project management hinges on several core responsibilities:

- **Stakeholder Management**: Project managers must ensure that all parties involved, from team members to clients, are aligned with the project's objectives and are informed of progress and challenges.

- **Risk Management**: It is crucial to identify potential risks to the project's timeline, budget, or quality and mitigate these risks.
- **Resource Allocation**: Efficiently distributing resources, whether human, technological, or financial, is critical to maintaining project efficiency and effectiveness.

These responsibilities form the backbone of project management, ensuring that projects are completed, meet the intended goals, and deliver significant value to the organization or client.

Impact of Effective Project Management

The impact of effective project management is not just significant; it's profound. Well-managed projects save time and costs, enhance customer satisfaction, and contribute positively to the organization's reputation. For instance, a study by the Project Management Institute (PMI) revealed that organizations that excel in project management waste 28 times less money than their counterparts that lack such practices. This inspiring statistic underscores the transformative power of effective project management, motivating you to strive for excellence in your project management journey.

Effective project management also facilitates better agility and adaptability in organizations, enabling them to respond more effectively to changes and opportunities. This adaptability is crucial in today's fast-paced business environments, where pivoting and embracing change can provide a significant competitive advantage.

In conclusion, understanding the fundamentals of project management, including the roles, responsibilities, and impacts, sets the stage for successful project execution. This knowledge forms the foundation upon which practical skills and strategies can be built, allowing you to manage projects successfully and confidently. As we progress through this book, you'll gain deeper insights into how these principles can be applied in various scenarios, preparing you to tackle your projects with expertise and assurance.

1.2 Demystifying Project Management Jargon: Key Terms Explained

In the intricate dance of project management, understanding the language is akin to knowing the steps in a complex choreography. Every term and acronym carries weight, shaping the interactions and decisions that drive a project forward. This section aims to clarify some of the fundamental jargon you encounter, providing a sturdy linguistic bridge from novice uncertainty to managerial fluency.

Project management is rife with specialized vocabulary that, when understood, can significantly enhance the clarity and efficiency of your projects. Let's start with some of the most common terms. The term '**scope**' refers to the project's boundaries, detailing what is to be accomplished. It defines the limits of the project deliverables and the breadth of work to be performed. For example, in a software development project, the scope might include developing a new application feature but not overhauling the existing user interface.

Milestones are another critical term, representing significant checkpoints or goals within a project's timeline. These are crucial for segmenting a project into manageable phases and for celebrating progress along the way. Think of milestones as lighthouses guiding a ship; they provide targets that, when achieved, affirm that your project is on the right course. In constructing a new building, a milestone might be the completion of the foundation, which is essential before any further construction can proceed.

The **Gantt chart**, named after its inventor, Henry Gantt, is a visual project management tool that illustrates a project schedule. It displays tasks over time, showing start and finish dates and dependencies between tasks. This tool is handy in projects where timing and coordination of tasks are critical. For instance, in an event planning project, a Gantt chart can help synchronize venue booking, catering services, and guest invitations, ensuring a seamless flow of preparatory steps leading up to the event.

Stakeholders are individuals or groups interested in a project's success or failure. They can range from team members and managers to external entities like clients, suppliers, and regulatory bodies. Effective stakeholder management involves identifying all parties affected by the project and engaging them appropriately to ensure their expectations are managed, and their contributions are aligned with project goals.

Deliverables are tangible or intangible outputs generated during the project. You commit to delivering These products or services to a client or stakeholder. In a marketing project, a deliverable might be a completed campaign strategy document, a series of advertisements, or an analytical report on campaign performance.

Methodology-Specific Terms

As you dig deeper into project management, you'll encounter terms specific to various methodologies that guide how projects are structured and executed. In Agile project management, for instance, '**sprints**' are short, consistent cycles in which teams complete workable chunks of a product. Think of a sprint as a mini-project, with its planning, execution, and review stages all condensed into a short period, usually two to four weeks.

The '**backlog**' in Agile is a list of tasks or features that must be completed. It is a to-do list of tasks assigned to specific sprints based on priority. Meanwhile, in the Waterfall methodology, a '**phase gate**' is a checkpoint at the end of a phase where work is reviewed and must be approved before moving on to the next phase. This is critical in projects where the sequential execution of phases ensures that foundational elements are approved and correctly completed before more dependent, detailed work begins.

Importance of Terminology in Communication

Grasping this terminology is not merely an academic exercise; it's essential for effective communication within your team and with stakeholders. Misunderstandings in project management can lead to misaligned expectations and errors in project delivery, jeopardizing the project's success. For example, if a team member needs to understand the project's scope, including an additional feature that stakeholders have not approved, it could lead to wasted resources and potential conflict.

Interactive Glossary

As you embark on your project management journey, consider the practicality of creating or utilizing an interactive glossary. This digital tool or document can serve as your quick reference guide, helping you understand and apply project management terms correctly in real-world scenarios. Reinforcing your understanding can make your journey into project management smoother and more confident. *I've also included a definition of the top 100 Project Management near the end of the book as a reference for your journey.*

1.3 Overview of Project Management Methodologies: Agile, Waterfall, and More

Navigating the diverse landscape of project management methodologies can often seem like exploring a vast and complex terrain, each path with its unique rules and tools. Understanding these methodologies is crucial, as each offers distinct advantages and challenges, and the choice of which to implement can dramatically affect the outcome of your projects. From Agile to Waterfall, SCRUM to Kanban, the evolution and diversity of these methodologies have shaped the project management field.

The Agile methodology, renowned for its flexibility and iterative approach, is a powerful tool for accommodating changes and frequent feedback. It breaks down projects into small, manageable units known as sprints, empowering teams to adjust their plans quickly in response to changing requirements. This method is particularly effective in environments where the end goal is partially defined, and the project must adapt. Industries like software development and digital marketing, where innovation and speed are crucial, find Agile especially beneficial.

In contrast, the Waterfall methodology offers a more structured approach, where each project phase must be completed before the next begins. This method is highly systematic, with clear milestones and deadlines set from the beginning. It works well in industries such as construction and manufacturing, where steps are linear, and changes during the execution phase can be costly. The predictability of Waterfall makes it easier for these industries to manage large-scale projects with stringent timelines and budget constraints.

SCRUM, often considered a subset of Agile, focuses heavily on team dynamics and emphasizes the importance of frequent updates, with short daily meetings and reviews at the end of each sprint. This framework fosters a highly collaborative environment, ideal for projects where team interaction can significantly influence outcomes. Software development teams, for example, thrive under SCRUM, allowing them to rapidly adjust to new insights and user feedback without disrupting the development flow.

Kanban, another Agile-based methodology, uses visual boards to manage work as it moves through various process stages. This method is fantastic for ongoing projects with continuous outputs, such as content production or ongoing maintenance work in IT services. It allows teams to see the status of each piece of work at any time, enhancing transparency and enabling better control over the workflow.

Comparative Analysis

Each of these methodologies carries its strengths and weaknesses. Agile is excellent for projects requiring flexibility and innovation, but it can be less predictable regarding budget and timeline. Waterfall offers predictability and straightforward planning, which can simplify resource allocation and long-term scheduling; however, its rigid structure makes it less adaptable to change. SCRUM excels in fostering team collaboration and rapid iteration but can require a significant shift in team dynamics and company culture to implement effectively. Kanban provides clarity and continuous delivery but might not suit projects needing definite phases and complex integrations.

Industry-Specific Recommendations

Choosing the proper methodology often depends on the industry and specific project requirements. Agile, SCRUM, and Kanban are predominant in software development because of the need for rapid development cycles and the ability to adapt to evolving technological landscapes. Waterfall is often preferred for construction projects due to its sequential execution, which aligns well with the linear nature of construction tasks. In healthcare, where compliance and regulatory requirements are critical, Waterfall provides the structure to ensure that each phase meets stringent standards before moving forward.

Methodology Adoption Tips

Adopting a new project management methodology within an organization involves more than understanding the theory; it requires a strategic approach to integration. Start by educating your team about the new methodology's benefits and challenges. Training sessions and workshops can be invaluable in this regard. It's also beneficial to pilot the methodology on a small project first, allowing your team to adjust to the new processes in a controlled environment. Feedback from these initial projects can help refine your approach before a full-scale rollout.

Change management, a crucial component in project management, plays a significant role in project success. It addresses resistance by emphasizing how the new methodology will benefit the team and the project outcomes. Regularly reviewing the adoption process and making adjustments based on team feedback and project performance is critical. This iterative approach to implementation, which mirrors the principles of Agile methodologies, ensures that the adoption is as smooth and effective as possible.

Mastering project management methodologies and their applications is a powerful tool that empowers you to make informed decisions about managing your projects. This understanding ensures that your chosen methodology aligns with your project's needs and team's capabilities. As you delve deeper into the functionalities and benefits of each methodology, you'll gain the confidence to lead your projects towards successful outcomes, harnessing the strengths of these diverse and robust project management tools.

1.4 Setting Up Your First Project: A Step-by-Step Guide

Initiating and managing your first project can be an exhilarating experience with opportunities and challenges. The key to a successful project lies in meticulous planning and execution, which gives you a sense of control and provides a security net. Let's walk through the foundational steps of setting up a project, from initiation to monitoring, ensuring you are well-equipped to take your project from concept to completion.

Project Initiation

The initiation phase is not just a step but the cornerstone of any project, setting the tone for all the following stages. At this crucial stage, it's essential to precisely define the project's scope, goals, and objectives. This will lay a clear roadmap of the project's goals and the parameters guiding its execution. For instance, if the project's mission is to launch a new software product, the scope should encompass all the critical phases from conceptual design and development to marketing and eventual market launch. Each phase should have specific objectives and anticipated deliverables, paving a clear path for the project's journey. Articulating these foundational elements with precision is of utmost importance.

A well-defined project scope safeguards against scope creep, where the requirements expand beyond the original parameters, often leading to cost overruns and delays. Equally important is setting objectives that adhere to the SMART criteria—specific, measurable, achievable, relevant, and time-bound. These criteria ensure that goals are clear and actionable, facilitating focused effort throughout the project's lifecycle and enhancing communication with stakeholders and team members. Establishing these project cornerstones with clarity and precision lays the foundation for a structured and successful project management process.

Planning Essentials

After initiating the project, we move into the planning phase, which is pivotal in project management. In this phase, you craft a comprehensive roadmap for reaching the project's goals. Begin by constructing a detailed project plan, which should encompass timelines, budget forecasts, and resource distribution strategies. Break the project into smaller, manageable tasks, each with its deadline. This breakdown can be effectively organized and visualized through Gantt charts. These tools offer a panoramic view of the project's schedule, facilitating easy progress tracking against established milestones. The cost estimates must be as precise as possible, covering every potential outlay from workforce to materials and equipment. Additionally, integrating a contingency fund is a prudent strategy. This fund is a financial cushion, enabling the project to navigate unforeseen challenges without compromising its integrity.

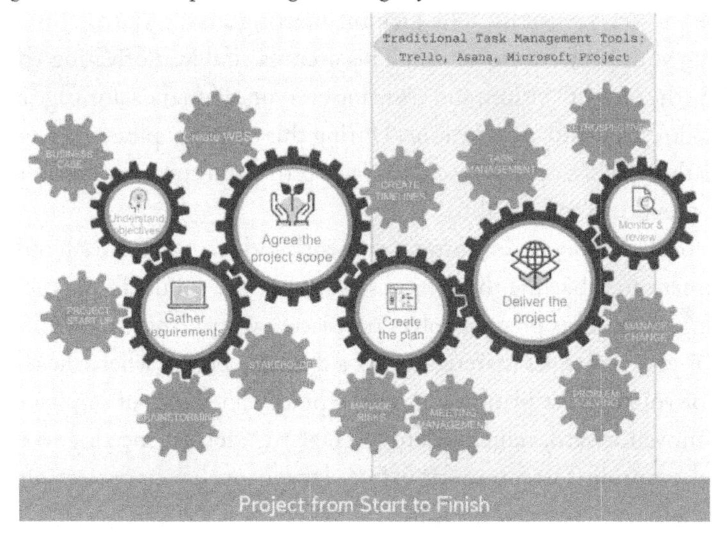

Next, we address resource allocation, a cornerstone of the planning phase. This step requires a careful assessment of the project's needs in terms of human, technological, and financial resources and their optimal distribution to ensure peak efficiency. For instance, assigning tasks to team members based on their specific skill sets guarantees that tasks are executed proficiently and enhances overall project efficiency. Similarly, ensuring the timely availability of technological tools and financial resources is crucial in averting potential delays and productivity bottlenecks. This detailed approach to planning lays a solid foundation for the subsequent phases of the project, positioning it for success. By meticulously outlining every aspect of the plan, from timelines and costs to resource allocation, you equip the project with the structure and clarity needed to navigate the complexities of execution and management.

Execution Tips

With your detailed plan, it's time to bring your project to life during the execution phase. The first crucial step is to energize and unite your team. It's essential that every team member clearly understands their specific roles, grasps the overarching project objectives, and recognizes the importance of their contributions to the project's success. Establishing a solid communication framework is critical; consider setting up regular, scheduled team meetings and making the most of digital communication tools to ensure a steady flow of information. This approach guarantees that every team member stays informed, aligned, and fully engaged with the project's progress. The kick-off meeting marks a pivotal moment in launching your project into action. This event is vital in motivating your team, clarifying the project's vision, and ensuring everyone is on the same page regarding strategic direction and expectations. During this meeting, please review the project plan thoroughly, openly discuss any concerns, and encourage team members to share their ideas and insights.

This collaborative process fosters teamwork and strengthens each individual's commitment to achieving the project's goals, laying a robust foundation for the upcoming work. This phase involves translating your plans into action and observing as your strategies materialize. It's a dynamic period where the theoretical aspects of your project planning meet the practical realities of implementation. As you move forward, remember to leverage the kick-off meeting to its fullest potential, ensuring it serves as a springboard that propels your project with clarity, purpose, and collective enthusiasm.

Monitoring and Adjusting

Once the project starts, it's critical to monitor it continuously to ensure everything aligns with the established plan. This phase entails diligently reviewing the project's progression and pinpointing any discrepancies between current outcomes and the original plan. By identifying these variances early, you are better positioned to implement necessary corrective measures, thereby averting potential derailments. The monitoring process is instrumental in promptly spotting challenges, allowing for adjustments that align the project's trajectory with its goals. The nature of these adjustments can vary significantly, from reallocating resources to better serve the project's needs to overhauling strategies when the current approach falls short. Alternatively, it might necessitate revisions to timelines and budgets to reflect the project's current reality. Mastering the art of adaptability is a quintessential skill for any project manager. For instance, should a pivotal task lag behind its scheduled completion, a project manager must assess whether deploying additional resources could hasten progress or if it's more prudent to recalibrate the timeline.

Regularly scheduled status updates and review meetings serve as vital mechanisms for this continuous appraisal, ensuring the project maintains its agility and capacity to respond to both internal dynamics and external pressures. Indeed, the journey from setting up to managing your first project is mapped out through a sequence of deliberate steps, starting from a well-articulated definition to comprehensive planning, followed by vigorous execution and rigorous monitoring. Each phase is intricately connected, creating a cohesive trajectory that steers the project toward its successful conclusion. As you navigate through these phases, remember that every project presents a unique learning curve, offering invaluable insights and opportunities to refine your project management expertise and grasp the complexities of this dynamic field. This journey is not just about learning; it's about empowering yourself with the skills and knowledge to lead successful projects.

1.5 The Importance of a Project Charter: Creation and Components

A project charter is not just a document but the very foundation of any project. Its importance cannot be overstated, as it is a comprehensive guide to the project's scope, objectives, and participants. Think of it as your project's constitution—a written document that outlines what the project entails and authorizes the project manager to initiate and lead it.

Charter Fundamentals

The essence of a project charter is to establish the framework and boundaries within which the project operates. It defines the project, why it is necessary, its goals, and who will be involved. A well-articulated project charter is critical as it is a contract between the project sponsor, key stakeholders, and the project team. It ensures that everyone involved understands from the outset, reducing the potential for conflicts or misunderstandings later on. For example, if a company wants to expand its operations by setting up a new office, the project charter would detail its purpose, scope, and resources, ensuring all stakeholders are aligned with its objectives.

Key Components

The project charter should encompass several vital components to be effective:

- **Project Purpose**: This clarifies why the project is necessary. It should align with the organization's broader strategic goals. For instance, the purpose of a project might be to increase production capacity by 30% to meet growing market demand.

- **Detailed Scope Description**: This outlines what the project will and will not cover, providing clear boundaries. It helps manage stakeholders' expectations and mitigates the risk of scope creep.

- **Budget**: The charter should provide an overview of the budget, including the money available and how it will be distributed across different project areas.

- **Key Stakeholders**: Identifying who has a stake in the project is crucial. The charter should list these parties and describe their roles and responsibilities.

- **Major Deliverables**: What the project aims to deliver should be clearly defined, with significant milestones and their expected completion dates.

Development Process

Developing a project charter is a collaborative effort that involves key stakeholders. This process typically begins with a series of meetings to discuss and define the project's purpose, scope, and objectives. It's crucial to involve all major stakeholders in these discussions to ensure their interests are understood and integrated into the project's planning process. For example, a software development project might include representatives from IT, marketing, customer service, and finance.

Each department's unique insights and requirements could impact the project scope and objectives. This collaborative approach ensures a comprehensive understanding of the project and makes every stakeholder feel included and valued in the project's development.

Once the initial information is gathered, the project manager drafts the charter and circulates it for feedback. Incorporating feedback is critical as it ensures that all perspectives are considered and that the final document has broad support. Once finalized, the charter must be formally approved by the project's sponsor, who is typically a senior executive or a representative of the organization's leadership. This approval is crucial as it grants the project manager the authority to move forward and allocate resources as outlined in the document.

Charter as a Communication Tool

More than just defining the project's framework, the charter is a powerful communication tool. It ensures all team members and stakeholders understand the project's goals and constraints. The charter is a reference point for project decisions about scope, resources, and priorities. For instance, if there is a request for an addition to the project deliverables, the project manager would refer to the charter to determine whether this change aligns with the project's defined scope and objectives. If the change is significant, it may require revising the charter and securing approval from the sponsor.

In practical terms, the project charter ensures that everyone involved is aligned, reducing the likelihood of conflicts and confusion as the project progresses. It acts as a north star, guiding the project through its phases, and provides a benchmark against which the project's success can be measured. Whether launching a minor initiative or a major enterprise-wide endeavor, creating a solid project charter is one of the first and most crucial steps toward achieving your project goals.

1.6 Essential Project Management Tools: A Beginner's Toolkit

In the vast landscape of project management, the tools you select can significantly shape the efficiency and triumph of your projects. These tools, ranging from sophisticated software to basic manual techniques, serve a specific purpose and cater to different project needs. Choosing the right tools is not merely about what is popular or advanced; it's about aligning the tool's capabilities with the project's requirements, team dynamics, and overall management style. This toolkit empowers you, the project manager, to select essential tools that perfectly match your project's size, complexity, and team nature, whether in one office or across the globe. By understanding the importance of tool selection and being equipped with the knowledge to make informed decisions, you can feel confident managing projects effectively.

The initial step in tool selection involves considering the project's scale and complexity. More straightforward tools best serve smaller projects with fewer tasks and stakeholders. On the other hand, larger, more intricate projects could benefit from more robust software that offers task management, resource allocation, and progress tracking features. The physical distribution of your team also plays a significant role. Tools that facilitate collaboration, communication, and real-time updates are indispensable for remote teams, showcasing the adaptability and reassurance these tools provide in various project scenarios.

Explore some popular project management software and understand its essential features and benefits. Trello, for example, uses an intuitive and visual card-based system. It's particularly effective for small—to medium-sized projects and teams that require flexibility. Each card can represent a task, and these can be moved across lists on a board to represent workflow stages. This setup is ideal for teams that benefit from seeing tasks and their statuses at a glance.

While both Asana and Trello are project management tools, they have distinct features that cater to different project management needs. Asana, for instance, offers more comprehensive project tracking. It organizes tasks in a shared workspace where team members can assign work, set deadlines, and comment on tasks. This software suits teams that need a bit more structure than Trello offers but still values simplicity and ease of use. Asana's capabilities make it easy to track dependencies and workflows, which are crucial for mid-sized projects with several interconnected tasks. The user-friendly interface of Asana and Trello ensures that even those new to project management tools can quickly adapt and start using them effectively.

Microsoft Project is a standout choice for those managing more significant, complex projects. It provides extensive features that help with scheduling, detailed task management, and resource allocation, all essential for large-scale projects. MS Project allows managers to draft a detailed plan while monitoring the project's progress and adjusting schedules as needed. Its ability to integrate with other Microsoft Office applications, such as Excel for data analysis and Word for project documentation, makes it a robust tool for organizations that rely on those tools for business operations. With Microsoft Project, you can feel confident in your ability to handle even the most complex projects, knowing that you have a powerful tool at your disposal.

In addition to digital tools, traditional project management instruments like Gantt charts and PERT diagrams still hold significant value. A Gantt chart, for example, offers a visual timeline for the project, showing when each task should start and finish, how long it will take, and overlapping activities. This tool is invaluable for planning and monitoring the project's progress and ensuring that critical milestones are met on time. On the other hand, PERT diagrams, which focus on the tasks required to complete a project and the time needed to complete each task, are excellent for identifying the minimum time necessary to complete a project and spotting potential bottlenecks. These tools provide a clear and visual representation of the project's timeline and tasks, making it easier for project managers to plan, monitor, and adjust their projects.

Integrating these tools into daily project activities can dramatically enhance your project's efficiency and your team's productivity. As the project manager, your role in this process is crucial. To effectively incorporate these tools, ensure all team members are trained to use them. This might involve scheduled training sessions, ongoing support, and regular check-ins to address any challenges. Encourage the team to use these tools not just as a reporting mechanism but as a way of making their everyday tasks more accessible and more structured. For example, you can schedule a daily stand-up meeting where team members update their tasks in the project management tool, discuss any challenges, and plan their work for the day. This way, the tool becomes integral to their daily workflow, enhancing their productivity and the project's efficiency. By empowering your team with these tools, you're giving them the control and capability to succeed.

Regularly updating your project management tools is crucial to reflect project scope or schedule changes. This practice helps foster communication and transparency within the team. For instance, updating tasks in Asana or moving cards in Trello should be part of the daily routine, not an afterthought. This habit ensures that everyone on the team knows the project's status and responsibilities, reducing the chances of tasks falling through the cracks. Regular updates also help keep the project management tools practical and up-to-date, ensuring they meet the project's needs and contribute to its success. By emphasizing the importance of regular updates, you can make the team feel responsible for maintaining the accuracy of project information, thereby enhancing the project's overall efficiency.

These tools can create a single source of truth for project information. In project management, a single source of truth refers to a central, authoritative source of information used and trusted by all project stakeholders. This practice helps avoid confusion and ensures all decisions are based on the most current and accurate data. Whether updating a Gantt chart to reflect a new deadline or adjusting resource allocations in MS Project, keeping your tools up-to-date ensures that your project management process is as dynamic and responsive as the projects you manage. It also ensures that all project stakeholders work with the same, accurate information, reducing the risk of miscommunication and misunderstandings.

By thoughtfully selecting and integrating these tools into your project management practices, you equip your team with the necessary resources to succeed. This enhances the efficiency of your processes and fosters a culture of transparency and continuous improvement within your team. As you progress with these tools, remember that they are not just mechanisms for tracking progress—they are instruments that empower you and your team to build pathways to success. By emphasizing the role of these tools in fostering a culture of transparency and continuous improvement, you can make the team feel motivated and part of a progressive environment, thereby enhancing their engagement and commitment to the project's success.

1.7 Beginners Cheat Sheet to Project Management

The terms are important to grasp as they build layer upon layer. This cheat sheet provides a quick foundation for beginners. It covers the most important aspects of project management without diving too deeply into complex terminology or methodologies, as you'll see later in the book. Remember, learning project management is a gradual process. As you become more comfortable with these basics, you can explore more advanced concepts and techniques to enhance your project management skills. It's a good idea to come back to this again after you've read the book!

Project Basics:

Project: A temporary endeavor to create a unique product, service, or result
Stakeholders: People or organizations affected by or interested in the project
Project lifecycle: Initiation, Planning, Execution, Monitoring & Controlling, Closing

Key Roles:

Project Manager: Leads the project team and is responsible for project success
Project Sponsor: Provides resources and support, champions the project
Team Members: Perform project tasks and contribute their expertise

Project Initiation:

Project Charter: Document authorizing the project and outlining high-level details
Scope: What the project will and won't deliver
Objectives: Specific, measurable goals the project aims to achieve

Planning:

Work Breakdown Structure (WBS): Hierarchical breakdown of project deliverables
Project Schedule: Timeline of tasks, milestones, and deadlines
Budget: Estimated costs and allocated resources for the project
Risk Management: Identifying, assessing, and planning responses to potential risks

Execution:

Task Assignment: Allocating work to team members

Resource Management: Ensuring team members have what they need to complete tasks

Communication: Keeping stakeholders informed of progress and issues

Monitoring & Controlling:

Progress Tracking: Comparing actual performance to planned performance

Change Management: Handling requests for changes to the project scope or plan

Quality Control: Ensuring deliverables meet required standards

Closing:

Deliverable Handover: Transferring project outputs to the client or end-users

Lessons Learned: Documenting what went well and areas for improvement

Project Archive: Storing project documents for future reference

Essential Tools:

Gantt Chart: Visual representation of project schedule

Kanban Board: Visual tool for managing workflow

Project Management Software: Tools like Microsoft Project, Trello, or Asana

Key Concepts:

Critical Path: The sequence of tasks that determine the project's minimum duration

Milestone: Significant point or event in the project

Deliverable: Any measurable, tangible outcome produced by the project

Communication Tips:

Regular Status Updates: Keep the team and stakeholders informed

Active Listening: Pay attention to team members' concerns and ideas

Clear Documentation: Maintain precise records of decisions and changes

Time Management:

Prioritization: Focus on the most critical and urgent tasks first

Time Estimation: Realistically assess how long tasks will take

Avoiding Scope Creep: Prevent uncontrolled expansion of project scope

Team Management:

Team Building: Foster a positive and collaborative team environment
Conflict Resolution: Address and resolve team conflicts promptly
Motivation: Keep team members engaged and committed to the project

Risk Management Basics:

Risk Identification: Recognize potential threats to the project
Risk Assessment: Evaluate the likelihood and impact of identified risks
Risk Mitigation: Develop strategies to reduce or eliminate risks

Quality Management:

Quality Planning: Determine quality standards for deliverables
Quality Assurance: Implement processes to meet quality standards
Quality Control: Monitor and record results of quality activities

Stakeholder Management:

Stakeholder Identification: Determine who is affected by or can affect the project
Stakeholder Analysis: Understand stakeholders' interests and influence
Stakeholder Engagement: Involve and communicate with stakeholders appropriately

Project Documentation:

Project Plan: Comprehensive document outlining how the project will be executed
Status Reports: Regular updates on project progress, issues, and next steps
Meeting Minutes: Records of discussions and decisions made in project meetings

Budgeting Basics:

Cost Estimation: Predicting the costs associated with project activities
Budget Tracking: Monitoring actual spending against the planned budget
Cost Control: Taking action to keep the project within budget

Agile Project Management:

Sprints: Short, time-boxed periods where specific work is completed
Daily Stand-ups: Brief daily team meetings to discuss progress and obstacles
Backlog: Prioritized list of features or requirements to be implemented

Project Closure:

Final Report: Document summarizing the project's performance and outcomes

Client Acceptance: Formal approval of project deliverables by the client

Team Recognition: Acknowledging and celebrating team contributions

Continuous Improvement:

Feedback Collection: Gathering input from team members and stakeholders

Performance Analysis: Reviewing project metrics and outcomes

Knowledge Sharing: Distributing lessons learned to benefit future projects

Chapter 2

From Theory to Practice

Navigating the transition from understanding project management theories to applying them in real-world scenarios can be exhilarating yet daunting. This chapter is designed to bridge that gap, providing practical tools and methodologies that transform theoretical knowledge into tangible success in project management. Here, we dig into crafting effective project plans—a fundamental skill for any aspiring project manager. It's your blueprint, your map; without it, navigating the complex terrain of project management can become overwhelming.

2.1 Crafting Effective Project Plans: Templates and Tips

Creating a robust project plan is akin to drafting a detailed recipe for a complex dish. To ensure the outcome is successful, it would be best to have the right ingredients (resources), clear instructions (processes), and the proper tools (templates). Let's explore how to build this recipe from the ground up.

Utilizing Templates

Project planning templates are invaluable tools in the project manager's toolkit. They provide a 'structured format' for organizing information. In project management, a structured format refers to a predefined layout or arrangement of information that helps organize and present data logically and systematically. This means the templates are designed to guide you in organizing your project's information logically and systematically, ensuring nothing is overlooked. Templates such as Gantt chart templates and project timeline templates offer visual representations of the project's timeline, detailing when and how tasks will be executed and by whom. For instance, a Gantt chart template can help you plot out the phases of an IT upgrade project, from initial data gathering to the final rollout, allowing you to see overlapping activities and adjust resources accordingly.

Think of these templates as empowering frameworks. Depending on the project's complexity and scope, you can modify these templates to suit your needs. Using templates also helps maintain consistency across projects within your organization, ensuring that all project plans adhere to a standard format and making it easier for stakeholders to understand and follow the project's progress.

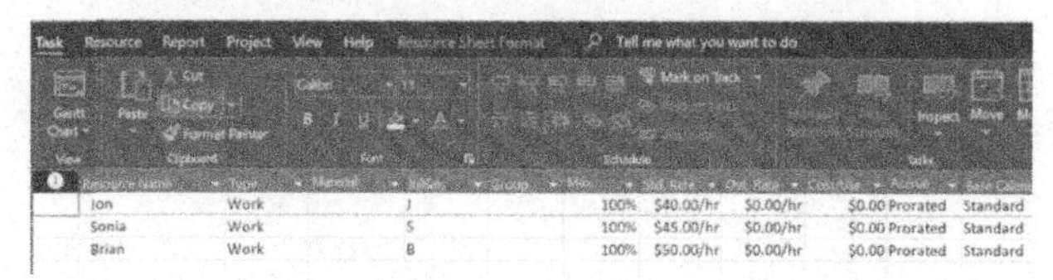

MS Project Template

Detailing the Planning Process

The planning process begins long before you fill in a template—it starts with gathering requirements. Engage with stakeholders to clearly understand what needs to be achieved. This can involve interviews, surveys, or brainstorming sessions. For instance, if you manage a software development project, you might need to gather requirements from the client, the end-users, and the development team. Once the requirements are precise, the next step is to break down the project into manageable tasks, a process known as the 'Work Breakdown Structure (WBS) '. This is a hierarchical project decomposition into smaller, more manageable components, tasks, or work packages. Each task is then assigned resources, deadlines, and responsibilities.

Setting realistic timelines and milestones is crucial. Use historical data from similar projects to inform your estimations and consider external factors that could impact your timeline, such as market conditions or regulatory changes. For example, if you are managing a construction project, you might need to adjust your timeline based on weather conditions or the availability of materials.

Incorporating Flexibility

Flexibility in project planning is not about losing control but about maintaining control even when unexpected changes occur. Techniques like buffer times or slack can be integrated into your schedule to absorb delays without affecting the overall timeline. Scenario planning is another effective strategy; by envisioning various possible future scenarios (e.g., best-case, worst-case, and most likely case), you can develop contingency plans to address them if they arise. This approach empowers you to steer the project even in the face of uncertainty, instilling a sense of confidence in your ability to handle any situation.

Imagine you are planning a software deployment project. By building in buffer times, you can accommodate potential delays in software testing without derailing the deployment phase. Similarly, scenario planning will not only prepare you but also instill a sense of confidence in handling situations like critical bugs or changes in software requirements smoothly.

Engaging Stakeholders in the Planning Phase

Stakeholder engagement is not just a courtesy but a strategic element of successful project planning. It refers to involving stakeholders in project decisions, activities, and planning. Involve key stakeholders early in the planning process to gain their buy-in and ensure their expectations are aligned with the project's trajectory. This collaborative approach fosters a sense of ownership among stakeholders, making them feel valued and integral to the project's success. It highlights the 'strategic value of their involvement, 'which means that their participation is not just beneficial but crucial for the success of the project. It enhances the quality of the project plan by incorporating diverse insights and expectations.

Effective engagement strategies include regular update meetings, shared access to project planning documents, or collaborative workshops. For example, when working on a new product development project, involving marketing, sales, and customer service teams in the planning phase can provide valuable insights into customer needs and market trends, significantly shaping the project's direction and outcomes.

By meticulously crafting your project plan with thoughtful consideration of templates, detailing processes, flexibility, and stakeholder engagement, you set the stage for a well-orchestrated project execution. Each step in this planning phase builds a firmer foundation for your project, ensuring that when unexpected challenges arise, your project remains robust and resilient, ready to adapt and succeed.

2.2 Real-World Case Study: Successful Project Launch in Tech

Launching a successful project in the fast-evolving tech industry, with its unique challenges of rapid technological advancements and shifting market demands, requires sharp project management skills and a keen adaptation to emerging technologies and market demands. Consider the case of a groundbreaking mobile app development project undertaken by a leading tech company aiming to enhance user engagement through personalized content. The project's objectives were clear from the outset: to develop an app that met the functional requirements and delivered a superior user experience tailored to individual preferences.

The project's scope was ambitious, and the challenges were manifold. However, the team's resilience and adaptability shone through. They integrated advanced machine learning algorithms to analyze user behavior and personalize content effectively while also maintaining user privacy and ensuring a scalable solution. Their ability to handle these challenges with grace and efficiency is truly inspiring.

Strategies Employed

The success of this project was mainly attributed to the strategic implementation of agile methodologies. Agile is an iterative approach to project management and software development that helps teams deliver value to their customers faster and with fewer challenges. The project team adopted this approach, enabling them to continuously incorporate feedback and make adjustments quickly. This flexibility was crucial in addressing technological challenges and evolving market trends. For example, during the development phases, user feedback highlighted the need for additional privacy controls not initially included in the project scope. The agile approach allowed the team to integrate these features without significant disruptions to the project timeline.

Innovation was central to the project. The team utilized cutting-edge technologies such as artificial intelligence (AI) to power the app's content personalization feature. This distinguished the project from competitors and significantly enhanced user engagement, as the app could learn and adapt to user preferences over time.

Strong leadership, a key pillar of successful project management, was pivotal in guiding the project to completion. The project leader, who possessed technical skills and excelled in communication and stakeholder management, ensured that the team remained motivated and aligned with the project goals, while keeping stakeholders informed and engaged throughout the project lifecycle.

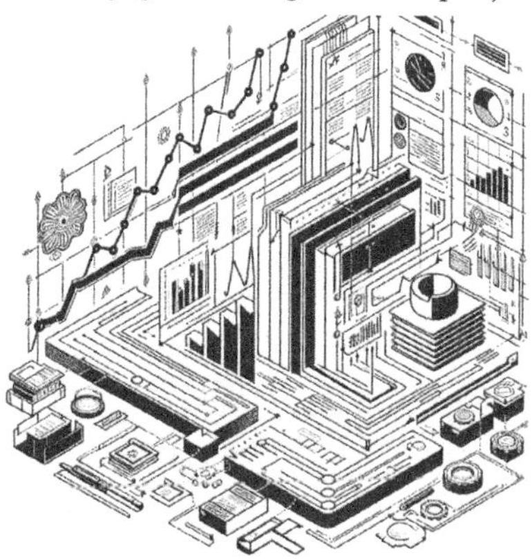

Challenges and Solutions

The integration of AI presented significant technical challenges, particularly regarding data privacy and security. The project team tackled these by adopting the 'best data encryption and anonymization practices. 'This refers to using advanced encryption algorithms and techniques to protect user data from unauthorized access. On the other hand, anonymization involves removing or modifying personally identifiable information from data sets, making it impossible to identify individuals. These practices ensured user data was protected in compliance with global privacy standards. Additionally, scalability was a concern as the app gained traction. To address this, the team developed a cloud-based solution that could dynamically scale resources based on user demand.

Another challenge was maintaining team productivity and cohesion, especially as deadlines approached and the workload increased. The project manager introduced regular check-ins and team-building activities that helped alleviate stress and foster a positive team environment. This proactive approach prevented burnout and kept the team focused and energized.

Lessons Learned

Lessons learned from this project can be valuable for anyone entering project management in the tech industry. Firstly, it's important to consider the significance of flexibility and adaptability in project management. Being able to pivot and adjust to new information and feedback was crucial in ensuring the project's relevance and effectiveness.

Secondly, integrating new technologies can greatly enhance project outcomes, although it can be challenging. However, thorough research and planning are necessary to ensure these technologies align with the project's objectives and are implemented effectively.

Lastly, strong leadership is crucial in navigating the complexities of tech projects. A leader who is technically proficient, skilled in communication, and has emotional intelligence can inspire and guide a team through challenging projects.

This case study shows that with the right strategies, tech projects can achieve their goals, set innovation and user engagement standards, and overcome challenges. As you manage your projects, consider these insights and techniques to enhance your approach and achieve success in your endeavors.

2.3 Implementing Agile Methodology in Your Project

Implementing Agile methodology in your projects can significantly enhance flexibility and responsiveness, which are crucial in today's fast-paced work environments. Agile, primarily known for its iterative approach and focus on continuous improvement, involves several key steps that you should follow to ensure successful adoption and execution.

Steps for Agile Adoption

The initial step in adopting Agile involves setting up your Agile team. This team should ideally consist of members who are adaptable, collaborative, and open to embracing change. Each member should clearly understand their roles within the team, which are often less hierarchical than in traditional project teams. This empowerment of team members, whose contributions are valued and integral to the project's success, is an essential aspect of Agile. Once the team is formed, the next crucial step is to define your sprints. Sprints are short, consistent periods—typically one to four weeks—during which specific project tasks are completed. Planning sessions at the start of each sprint allow team members to discuss and agree upon the work to be done. Following this, daily scrum meetings, which are brief stand-up meetings, should be held to discuss progress and any immediate issues that need addressing. These meetings are pivotal as they foster open communication and prompt problem-solving.

Organizing these scrum meetings effectively is critical. They should be kept concise and focused, with each team member answering three core questions: What did I accomplish yesterday? What will I do today? What obstacles are impeding my progress? This format, which emphasizes the collaborative nature of Agile, helps keep the team aligned and quickly identifies any hurdles that could disrupt the sprint, allowing for swift resolution. This fosters a sense of teamwork and camaraderie among the audience, making them feel more connected and involved in the project.

Tools for Agile Management

Agile management tools and software solutions are not just theoretical concepts but practical aids that significantly enhance the effectiveness of Agile projects. They are about collaboration, tracking progress, and real-time visibility into each team member's tasks and progress. Take Jira, for example, a popular tool among Agile teams. It's not just a tool but a platform that offers accessible features for planning, tracking, and managing Agile software development projects. It allows teams to create user stories, plan sprints, and distribute tasks across the team. Similarly, Asana is not just a task management tool but a comprehensive solution that provides task assignments, deadlines, and progress updates. These tools are critical and indispensable as they provide visibility into each team member's tasks and progress, ensuring everyone is aligned with the current priorities and deadlines.

The benefits of using such tools are manifold. They help maintain a clear overview of the project's progress and can significantly enhance productivity by automating various aspects of project management, such as sprint planning and performance tracking. Additionally, they encourage transparency and accountability, as team members can see real-time updates, fostering a culture of openness and collective responsibility.

Common Pitfalls and How to Avoid Them

Several common pitfalls can undermine the effectiveness of Agile. One typical mistake is not fully committing to the Agile process. Some teams may implement Agile practices while clinging to traditional project management methods, leading to clarity and efficiency. To avoid this, ensure that all team members and stakeholders understand Agile principles and their reasons for adopting them. Comprehensive training sessions and regular review meetings can help reinforce this understanding and commitment.

Another frequent issue impacting Agile projects is neglecting team dynamics and failing to foster effective communication. Agile doesn't just rely heavily on robust and collaborative team interactions but thrives on them. Refraining from addressing issues such as poor communication, lack of trust, or failure to involve all team members actively can lead to project delays and morale problems and undermine Agile's essence. Facilitating team-building activities and providing communication tools are not just suggestions but essential steps that can help create a supportive and collaborative team environment.

Measuring Agile Success

Specific metrics and Key Performance Indicators (KPIs) should be established to assess the success of Agile implementation in your projects. Velocity, a standard Agile metric, measures the work a team completes during a sprint. It helps predict the team's future performance and plan future sprints more accurately. Another helpful metric is the sprint burndown chart, which tracks the amount of work remaining in a sprint day by day. This can be an excellent tool for visualizing daily progress and ensuring the team is on track to complete the sprint goals.

Lead time and cycle time are also essential KPIs in Agile projects. Lead time measures the total time taken from the request of a new feature until its delivery, while cycle time measures the time taken to complete a task. Monitoring these times can help identify bottlenecks and improve the process's efficiency.

Implementing these metrics allows you to assess how well Agile is being adopted quantitatively and its impact on project outcomes. Regularly reviewing these metrics provides insights into areas for improvement, ensuring that your Agile practices continue to evolve and contribute positively to project success. By effectively setting up Agile teams, utilizing the right tools, avoiding common pitfalls, and measuring success with specific metrics, you can fully leverage the benefits of Agile methodology, leading to more successful and adaptable project management.

2.4 Risk Assessment and Mitigation Strategies

In the dynamic landscape of project management, risks are inevitable companions to every endeavor, whether launching a new product line, upgrading IT systems, or implementing a marketing campaign. However, by understanding how to effectively identify, evaluate, and mitigate these risks, you can proactively safeguard your project's outcomes and enhance your capability as a foresighted and prepared project manager. This emphasis on proactive risk management can instill a sense of preparedness and ability to handle potential challenges, making you feel more secure and capable in your role.

Identifying Risks

The first step in managing risks is to identify them accurately and comprehensively. Techniques such as SWOT analysis (Strengths, Weaknesses, Opportunities, Threats) and risk mapping are efficient tools for this task. SWOT analysis helps you view the project from various angles, pinpointing potential risks and identifying opportunities that could be leveraged. For instance, while the 'Weaknesses' and 'Threats' sections help you spot the risks, 'Strengths' and 'Opportunities' can offer insights into mitigating or turning challenges into advantages.

On the other hand, risk mapping involves visualizing the potential risks identified in a project. This could be as simple as a two-axis table where one axis represents the likelihood of the risk, and the other represents the impact. Placing risks into this matrix provides a clear picture of which risks need immediate attention and could be monitored over time. Still, it also enhances your ability to communicate these risks to stakeholders. This method simplifies the risk analysis process and ensures that stakeholders are aware of and prepared for potential challenges, boosting your confidence in your ability to manage and communicate risks effectively.

Evaluating Risk Impact

Once risks are identified, the next crucial step is to evaluate their potential impact on the project. This evaluation should consider both the likelihood of the risk occurring and the severity of its effect if it does happen. Prioritizing risks based on these factors helps you to allocate resources and attention where they are most needed. For example, a high-impact, high-likelihood risk, such as losing a key supplier, requires a more immediate and robust mitigation strategy than a low-impact, low-likelihood risk, like a minor delay in a non-critical task.

Practical risk evaluation also involves continuous monitoring, as new and existing risks can emerge over the project's duration. Engaging team members in regular risk assessment meetings encourages a culture of vigilance and proactive management, ensuring that risks are identified and addressed promptly.

Developing Risk Mitigation Plans

Developing comprehensive risk mitigation plans is essential for managing your identified and evaluated risks. These plans should include strategies for avoiding the risk, reducing its impact, transferring it to another party, or accepting the risk if it falls within the project's tolerance levels. For instance, in the case of potential vendor delays, you might mitigate the risk by selecting multiple suppliers or having a contingency plan that includes buffer stocks.

Contingency planning involves identifying alternative plans of action that can be implemented if a risk materializes. This is crucial for ensuring your project remains on track despite unforeseen challenges. Risk transfer, another effective strategy, might involve outsourcing specific tasks to third-party vendors who can assume the risk under contractual agreements that protect your project's interests.

Regular Review and Adjustment

Risk management is not a one-time task but a continuous process that requires regular review and adjustment. As your project progresses, the risk landscape can change significantly, necessitating updates to your risk assessment and mitigation strategies. Regular risk review meetings involving key project team members and stakeholders should be scheduled. During these meetings, discuss any new risks that have emerged, review the status of risks on the risk register, and adjust mitigation strategies as necessary.

This ongoing process helps manage risks more effectively and builds resilience within the team, enabling them to handle challenges more adeptly. Adjustments might include reallocating resources to areas with newly identified risks or scaling back mitigation measures for risks that have become less relevant. This dynamic approach ensures that your project adapts to changing conditions, maintaining momentum toward its objectives while safeguarding against setbacks.

By embedding a robust risk management process into your project management practices, you equip yourself and your team with the knowledge and tools to anticipate, address, and navigate uncertainties. This proactive stance on risk management protects your projects from potential derailments. It enhances your credibility and effectiveness as a project manager, ready to tackle challenges head-on and steer projects to successful completion.

2.5 The Art of Effective Communication in Projects

Effective communication stands as the backbone of successful project management. It ensures that every team member and stakeholder is on the same page, which is crucial for the smooth execution and timely delivery of projects. Creating a comprehensive communication plan is your first step toward establishing a clear communication framework. This plan serves as a roadmap, detailing who needs to receive what information, when they need it, and how it will be delivered. For instance, consider a scenario where you need to update stakeholders on project progress. Your communication plan should specify the frequency of these updates, the format (e.g., email, dashboard), and the level of detail required. It should also identify the primary point of contact for each type of communication to streamline the process and avoid overlaps.

When crafting this plan, start by listing all project stakeholders and team members, recognizing their unique needs and influence on the project. Categorize them based on these factors. Next, determine the types of information to be shared, ranging from status updates to decision requests, and the appropriate timing for these communications. For example, high-level stakeholders may only require monthly progress reports, while the project team might need daily briefs. The communication medium also plays a critical role. At the same time, email might suffice for sharing general updates, but complex issues may require face-to-face meetings or video conferences to ensure clarity and immediate feedback.

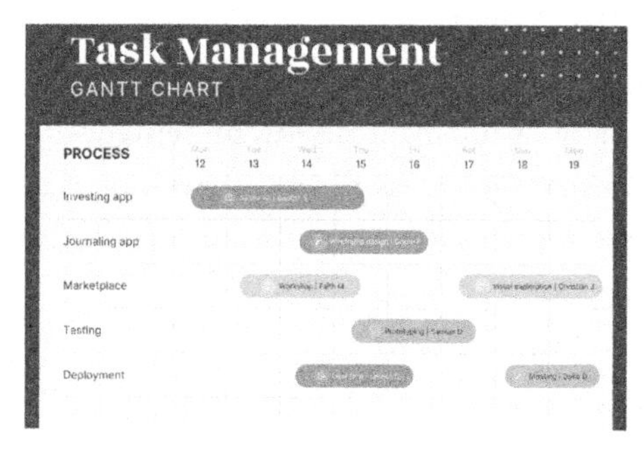

The choice of tools and platforms is a critical decision that significantly affects the efficacy of communication within project teams and with external stakeholders. Numerous digital tools facilitate seamless communication and collaboration. Tools like Slack or Microsoft Teams enable real-time messaging and document sharing, making them ideal for teams that require ongoing interaction. For more structured communications, such as project updates or milestone reviews, platforms like Zoom or Microsoft Teams meetings provide a space for detailed discussions and presentations. Project management software with built-in communication features, such as Asana or Monday.com, can be handy. These platforms integrate task management with communication, allowing team members to discuss tasks directly within the project's workflow, which enhances transparency and keeps all pertinent information easily accessible.

However, challenges can arise even with a solid communication plan and the best tools. Cultural differences, remote team dynamics, and simple miscommunications are common issues that can disrupt the flow of information. Understanding and respecting each team member's cultural background and communication preferences is essential to managing cultural differences. This might involve accommodating different time zones, languages, or communication styles. For remote teams, regular virtual check-ins can help maintain a sense of connection and ensure alignment. Establishing clear protocols for communication, such as regular status updates and feedback loops, can help mitigate misunderstandings and keep everyone informed.

Establishing robust feedback mechanisms is critical to ensure effective communication and continuous improvement in project management processes. Feedback mechanisms allow team members and stakeholders to express concerns, suggest improvements, and affirm when a communication strategy works well. This could be as simple as a suggestion box, regular feedback surveys, or a more structured approach like review meetings at key project milestones. The key is to make these feedback mechanisms easily accessible and to encourage honest, constructive communication. For example, after a project update meeting, you could send a follow-up survey to collect feedback on the clarity of the information presented, the effectiveness of the communication format, and areas for improvement. This not only helps in refining communication strategies but also fosters a culture of openness and continuous enhancement.

By meticulously planning your communication strategies, choosing the right tools, tackling communication challenges head-on, and establishing a culture of feedback, you create an environment where information flows smoothly, and every project member can effectively contribute to the project's success. This leads to better project outcomes and enhances team cohesion and stakeholder satisfaction, reinforcing the foundational role of effective communication in successful project management.

2.6 Monitoring and Controlling: Keeping Your Project on Track

Effective project management is not just about planning and execution; it's equally about consistent monitoring and controlling to ensure that the project remains aligned with your objectives and expectations. Setting up robust monitoring systems helps you keep a keen eye on the project's progress and swiftly address any deviations from the plan.

Setting Up Monitoring Systems

Establishing effective monitoring systems is vital to keeping your project on track. These systems are designed to provide continuous oversight of project activities, helping you detect any discrepancies between the planned and actual progress. Implementing tools like dashboards and progress reports is highly beneficial. Dashboards provide a real-time visual representation of key project metrics, allowing you and your stakeholders to understand the project's current status quickly. These might include metrics on project timelines, budget expenditures, resource usage, and more. For example, a dashboard could show that while the project is on budget, it is behind schedule, prompting immediate attention.

Progress reports complement dashboards by delivering a more detailed project status analysis. These reports should be generated at regular intervals—weekly, bi-weekly, or monthly—depending on the project's duration and complexity. Each report should provide insights into what has been accomplished, what is lagging, and any issues that need addressing. This continual flow of information is crucial for making informed decisions that align the project with its timelines and goals.

Controlling Scope Creep

Scope creep, the bane of many projects, refers to uncontrolled changes or continuous growth in a project's scope, often resulting in cost overruns or delays. To manage scope creep effectively, it's essential to have a transparent change management process in place. This process should outline how changes in scope are handled, including who has the authority to approve these changes and how they are documented and communicated.

Handling change requests involves assessing each request's impact on the project's budget, timeline, and resources. For instance, if a new feature request in a software development project could delay the launch, you need to weigh the benefits of the feature against the risks of postponing the product's release. Maintaining strict control over how changes are integrated into the project plan can prevent scope creep from derailing your project.

Quality Control Processes

Implementing quality control processes in project management is not just a necessity; it's a powerful tool that empowers you to take control and ensure that every output meets the required standards and, importantly, satisfies stakeholders' expectations. These processes, which involve regular quality checks at various stages of the project, equip you to maintain high quality in deliverables and prevent any deviations from the set standards. For instance, in a construction project, quality control might include compliance checks with building codes, regular inspections, and testing materials to ensure durability.

Quality assurance activities, such as peer reviews and testing, are not just crucial; they are your secret weapon. In software development, code reviews and beta testing are like your trusty sidekicks, helping you identify and fix bugs before the software is released and ensuring the product is functional and user-friendly. By embedding these quality control practices into your project management routine, you're not just preventing issues but ensuring that your projects are of the highest quality, avoiding costly reworks or project delays.

Conducting Project Reviews

Regular project reviews are critical for assessing whether the project remains on track and aligned with its objectives. These reviews allow the project team and stakeholders to reflect on what has been achieved, identify areas for improvement, and adjust strategies as needed. Reviewing the project's scope, budget, timelines, and quality standards is essential during these sessions.

Effective project reviews are not just structured; they're a platform for collaboration. They involve key project team members and stakeholders, allowing open dialogue and collaborative problem-solving. For example, if a review reveals that the project spends too much time on specific tasks, it's not just a problem; it's an opportunity. The team, including you, can explore ways to streamline these processes or allocate additional resources.

When conducted effectively, project reviews keep the project on track and foster a sense of accountability and transparency among team members. They serve as a platform for continuous improvement, ensuring the project can adapt to internal and external changes while still meeting its objectives.

By effectively setting up monitoring systems, controlling scope creep, implementing quality control processes, and conducting regular project reviews, you ensure that your project not only stays on track but also adheres to the highest standards of quality and efficiency. These practices are essential for navigating the complexities of project management and leading your projects to completion.

In wrapping up this chapter, we've explored various strategies to monitor and control your projects effectively. These methodologies serve as navigational tools, helping you steer your projects through unexpected challenges and changes. As we move forward, remember that the ability to adapt and respond to new information is just as critical as the initial planning and setup. In the next chapter, we'll dig deeper into advanced project management strategies, where you'll learn to elevate your skills and tackle even more complex project scenarios with confidence and expertise.

Chapter 3

Advanced Project Management Strategies

Let's not just venture further into project management; let's equip ourselves with advanced strategies that tackle complex project demands. This chapter is not just designed to transform your foundational skills; it's designed to empower you with a finely tuned ability to manage projects with a higher degree of sophistication and effectiveness. Here, we dig into resource allocation, a critical area where your strategic planning meets on-the-ground execution, giving you the tools to handle any project challenge.

3.1 Advanced Resource Allocation Techniques

Optimizing Resource Utilization

In the complex tapestry of project management, efficient resource utilization is akin to weaving together diverse threads to create a cohesive and functional whole. This requires a keen understanding of resource leveling and smoothing, two pivotal techniques that optimize available resources without overburdening your project infrastructure.

Resource leveling is a technique for balancing resource demand by adjusting the project schedule. It is beneficial when resources are limited or are required to work on multiple tasks or projects simultaneously. For instance, if you have a key team member who is overallocated, you can use resource leveling to delay specific tasks until the resource is available. This ensures that no single resource becomes a bottleneck, potentially derailing the project's progress.

On the other hand, resource smoothing is used when the project timeline is fixed, but resources are utilized flexibly. This technique adjusts the resources to spread more evenly across the tasks, ensuring that all functions receive adequate attention without any significant peaks or troughs in resource allocation. This can be particularly useful in maintaining a steady workflow and reducing downtime or periods of excessive workload.

Both techniques require a thorough understanding of the project's tasks, dependencies, and resource capabilities. They also necessitate continuous monitoring and adjustment to respond to project dynamics and resource availability, ensuring that the project remains on track while optimally using every resource.

Tools and Software for Resource Management

Leveraging advanced tools and software, such as SAP ERP or Jira, is crucial in managing resources effectively, especially in complex or large-scale projects. These Enterprise Resource Planning (ERP) systems and specialized project management software provide robust solutions for resource allocation, offering features that help you plan, monitor, and adjust resources as needed.

ERP systems integrate functions like project management, human resources, and finance into a unified system. This integration allows for a holistic view of resource availability and utilization across the organization, facilitating more informed decision-making. For example, if a project requires additional resources, the ERP system can quickly provide visibility into which departments have available capacity.

Dedicated project management software, such as Microsoft Project or Oracle Primavera, is tailor-made to handle complex project schedules and resources. These tools offer sophisticated functionalities like automatic resource leveling, allocating resources evenly over time, and avoiding overloading or underutilization. Real-time resource allocation dashboards and detailed reports are also provided, helping project managers optimize resource use, anticipate shortages, and adjust plans proactively.

Handling Resource Shortages

Resource shortages are a common challenge in project management, often leading to delays and increased costs. Strategic reallocation, prioritization of project activities, and outsourcing are effective tactics to mitigate these issues.

Reallocation involves shifting resources from lower-priority tasks to critical ones, ensuring that essential project milestones are met. Prioritization requires a clear understanding of project goals and deliverables, enabling you to make informed decisions about which tasks warrant more immediate resource investment.

Outsourcing is another viable solution when internal resources are insufficient. By contracting external vendors or freelancers for specific tasks or functions, you can fill gaps in your resource pool without the long-term costs of hiring new employees. For instance, in a software development project, you might outsource the design phase to a specialized design agency, ensuring high-quality design without hiring a full-time designer. However, managing these external resources carefully ensures they meet your project's quality standards and integrate smoothly with your internal teams.

Case Studies on Resource Allocation

Real-world case studies illuminate the impact of effective resource allocation on project outcomes. Consider the case of a technology firm that implemented resource smoothing. This technique involves redistributing workloads to even out resource utilization to manage the development of a new software product. Initially, the project experienced significant delays due to uneven resource distribution, which caused bottlenecks in critical development phases. By applying resource smoothing, the firm could redistribute workloads more evenly, ultimately reducing the development time by 20% and increasing team satisfaction due to a more balanced workload.

Another example involves a construction company facing resource shortages mid-project due to unexpected resignations. Using an ERP system to quickly assess resource availability and reallocate personnel from less critical tasks, the company could maintain the project timeline without additional hiring costs. Additionally, they outsourced specific specialized tasks to local contractors, ensuring project milestones were met on time.

These cases exemplify how strategic resource allocation can influence project efficiency and success. But there's more to project management than just resources and schedules. Emotional intelligence, an essential skill for project managers, can significantly impact project outcomes. By understanding and implementing these advanced techniques, you can navigate resource challenges more effectively, ensuring that your projects are completed within scope and budget and achieve the highest standards of quality and efficiency.

3.2 Mastering Scope Management to Avoid Creep

Defining and solidifying the project scope from the get-go must be addressed in project management. It acts as your project's blueprint, outlining what will be done and what will not. A well-defined project scope sets clear expectations for all stakeholders and provides a baseline against which project performance can be measured. To ensure everyone involved understands and agrees with the scope, it is crucial to document every detail meticulously. This documentation should include the project's objectives, deliverables, boundaries, and the responsibilities of each team member. Consider it a contract that all parties review and agree to; it clarifies the project's boundaries and reduces the likelihood of 'scope creep '—unauthorized changes or continuous growth in a project's scope that can lead to project delays, budget overruns, and stakeholder dissatisfaction.

One effective method to achieve a unified stakeholder understanding is to conduct interactive scope definition sessions. These sessions can be workshops where stakeholders collaborate to define the project scope. This collaborative approach ensures that all relevant insights and expectations are considered and fosters buy-in and commitment from everyone involved. For instance, in a software development project, involving end-users, developers, and business analysts in scope definition can help identify all functional and non-functional requirements, thus ensuring a comprehensive understanding of the project scope.

Once the scope is defined, maintaining its integrity becomes the next challenge. Regular scope reviews ensure the project remains on track and within its predefined boundaries. These reviews should be scheduled at key milestones throughout the project and involve revisiting the scope document to check if the project is progressing without deviation. Employing scope baselines is another crucial technique in scope verification. A scope baseline is a snapshot of the project's scope at a given time, which can be used to measure any changes or drift. This baseline, once established, serves as a standard to which you can compare the current project scope during reviews, helping identify any deviations early on.

Managing scope changes effectively is a critical aspect of scope management. No matter how well you plan, changes are inevitable. To handle these effectively, a structured change control system is necessary. This system outlines the process for handling change requests, including how changes are submitted, reviewed, approved, or rejected. It also details the roles of those involved in the decision-making process. A change control board, comprising key stakeholders who can assess the impact of proposed changes on project objectives and timelines, can be invaluable. For example, if a change request involves adding new features to a software product, the change control board would evaluate how these additions would affect the schedule, costs, and resource allocation, ensuring the project remains viable.

Preventive measures for managing scope creep involve proactive engagement and clear communication channels. Continuously engaging stakeholders throughout the project lifecycle ensures that all parties remain aligned to the original scope and that any emerging issues can be addressed promptly. Regular update meetings, clear documentation of agreed changes, and transparent communication help maintain this alignment. Additionally, educating your team on the importance of adhering to the defined scope and training them to recognize and handle scope creep can mitigate risks associated with unauthorized changes. Other preventive measures include setting realistic project goals, conducting regular scope reviews, and implementing a structured change control system.

By mastering these scope management techniques, you not only protect your project from the perils of scope creep but also instill a sense of empowerment. You ensure your project delivers the intended value to stakeholders, staying aligned with its original goals and objectives. This proactive approach to scope management is critical to maintaining control over your projects and ensuring they are delivered within the defined scope, time, and budget, giving you the confidence to handle any project challenges.

3.3 Leveraging Leadership Skills for Project Success

Navigating the multifaceted terrain of project management, leadership emerges as a cornerstone, pivotal for steering projects toward their successful completion. Leadership within project management transcends the conventional act of managing tasks; it involves inspiring, motivating, and influencing teams to achieve project goals. Understanding different leadership styles—transformational, transactional, and servant leadership—and their application based on project type and team dynamics is not just essential. It also enlightens you, making you more knowledgeable and effective as a project manager.

Transformational leadership, characterized by the ability to inspire and motivate team members, is invaluable in projects requiring a high degree of innovation and change. A transformational leader focuses on transforming others to help each other, to look out for each other, to be encouraging and harmonious, and to look out for the organization as a whole. For instance, in a start-up launching a new technology, a transformational leader might energize the team by setting a compelling vision, fostering an environment of creativity, and encouraging risk-taking with a focus on long-term success.

In contrast, transactional leadership, which focuses on maintaining the normal flow of operations, can be effective in projects with precise, structured tasks and established procedures. This style hinges on the principle of rewards and penalties based on performance results. It is instrumental in construction projects or manufacturing, where tasks are repetitive and emphasizes efficiency and adherence to proven procedures. Here, the leader may focus on setting clear objectives, monitoring performance, and providing direct feedback to keep the project on track.

Servant leadership, which emphasizes the growth and well-being of team members and the communities to which they belong, can be particularly effective in projects that require strong teamwork and high stakeholder engagement. Servant leaders put their employees first, understand their personal needs and interests, and develop them into more potent employees and people. In a community development project, a servant leader might focus on building trust and fostering collaborative relationships within the team and with the community members impacted by the project.

Each leadership style has strengths and can be effective in different project contexts. Choosing the right style requires understanding the project's specific needs and the team's dynamics. Adapting aspects of multiple leadership styles to address best the unique challenges and opportunities each project presents is often beneficial.

Influencing and Motivating Teams

Beyond selecting the appropriate leadership style, the ability to influence and motivate a team is crucial for driving project success. Compelling team motivation involves understanding and aligning with their values, aspirations, and needs. Techniques for building team cohesion and morale are critical, especially in long-term projects where motivation can wane.

One effective strategy is establishing clear, attainable goals aligned with team members' growth objectives. This enhances productivity and ensures team members are engaged and invested in the project's success. Regular recognition of big and small achievements also plays a critical role in sustaining motivation. Whether through formal recognition programs, celebratory team events, or simple verbal acknowledgment during meetings, recognizing individual and team contributions boosts morale and makes your team feel appreciated and valued, encouraging them to continue delivering excellence.

Creating a transparency and open communication culture is another crucial factor in motivating teams. When team members feel informed about the project's progress and challenges and their input is valued in decision-making, they are more likely to be committed and proactive. Regular team meetings, open forums for sharing ideas and concerns, and consistent updates on project status and developments can foster an environment of trust and inclusivity.

Decision-Making in Project Management

Decision-making in project management is often complex, involving multiple stakeholders and a wide array of variables. Influential project leaders are adept at making informed, timely decisions that propel the project forward while managing risks and uncertainties. This requires a deep understanding of the project's objectives, the needs and dynamics of the team, and stakeholders' expectations.

One crucial technique to bolster decision-making in complex projects is to foster a collaborative approach, involving key team members in the process. This brings diverse perspectives and expertise to the table and enhances the quality of decisions. Additionally, it ensures buy-in from those responsible for implementing them, fostering a sense of ownership and commitment.

During project crises, the role of emotional intelligence becomes even more pronounced. Influential leaders with high emotional intelligence rely on their experience, intuition, and a clear understanding of the project's priorities to make quick and effective decisions under pressure. The ability to remain calm, gather and analyze information swiftly, and communicate decisions clearly to the team is a testament to the power of emotional intelligence in crisis management.

Leadership Challenges and Solutions

Managing diverse teams is a challenge that can be effectively addressed through empathy. A key leadership strategy involves recognizing and respecting individual differences and finding ways to turn diversity into a project strength. This might involve tailored communication strategies, culturally sensitive management practices, or team-building activities that celebrate and leverage diversity. The value of empathy in this context cannot be overstated, as it helps in understanding individual differences and leveraging them for project success.

Leading without formal authority, a common scenario in matrix organizations or projects where team members do not directly report to the project manager can be particularly challenging. In such cases, building credibility and trust becomes paramount. This can be achieved through demonstrating competence, reliability, and fairness in interactions with the team. Developing strong interpersonal relationships and engaging team members in decision-making can help build influence and lead effectively, even without formal authority.

By mastering these advanced leadership skills and applying them in various project scenarios, you can enhance your effectiveness as a project manager and empower yourself to drive your projects to success. This empowerment fosters an environment that promotes growth, collaboration, and achievement among your team members, instilling confidence in your abilities.

3.4 Emotional Intelligence in Project Management

Fundamentals of Emotional Intelligence

Navigating project management's complexities requires more than technical know-how and organizational skills; it demands high emotional intelligence (EI). Emotional intelligence is the ability to perceive, control, and evaluate emotions—both your own and those of others. It comprises five core components: self-awareness, self-regulation, motivation, empathy, and social skills. Each plays a critical role in enhancing your effectiveness as a project manager.

Self-awareness is the cornerstone of EI. It involves understanding your own emotions, strengths, weaknesses, and the impact of your actions on others. This awareness is crucial in managing your behavior and making informed decisions. For instance, recognizing when you are stressed and understanding how it affects your decision-making can help you take steps to mitigate its impacts. Encouraging self-reflection can foster a deeper understanding of your emotions and actions, leading to more effective management.

Self-regulation relates to managing emotions healthily, maintaining control and adaptability, and staying committed to personal accountability. This skill is essential in project management, where pressure and stress are standard. Staying calm and composed during challenging situations helps maintain team morale and keeps the project on track.

A passion for work beyond money or status characterizes motivation in emotional intelligence. Motivated leaders are driven to achieve accomplishment, not just external rewards. This intrinsic motivation can inspire and influence team members positively, driving the collective project effort toward success.

Empathy, or understanding and sharing another person's feelings, is invaluable in managing teams and stakeholders. It allows you to navigate personal interactions more effectively and to resolve conflicts more amicably. By understanding the perspectives and emotions of team members, you can address concerns more effectively and foster a more cooperative and supportive project environment.

Finally, social skills in EI involve managing relationships to move people in desired directions, whether leading or negotiating. These skills enable you to build and maintain effective teams, facilitate better communication, and create an atmosphere that promotes productivity and satisfaction.

Applying EI to Project Management

Emotional intelligence (EI) is not just a tool but a transformative superpower that equips project managers, team leaders, and other professionals to revolutionize their interactions and relationships with team members and stakeholders. It's a game-changer in leadership, conflict resolution, teamwork, and project success. By harnessing EI, you can transform team communication, boost engagement, and enhance efficiency, inspiring you to steer your projects toward triumph confidently.

For instance, a project manager with high emotional intelligence might notice signs of burnout or frustration in the team, such as decreased productivity or increased absenteeism. They could then take proactive steps to address these issues before they impact project performance. This might involve adjusting project workloads, providing additional support, or offering encouragement and recognition of hard work.

Moreover, EI is not just important; it's a secret weapon in stakeholder management—a pivotal component of successful project delivery. Understanding and empathizing with stakeholder concerns and expectations can be a game-changer, leading to more effective communication and negotiation. This, in turn, can help align project goals with stakeholder needs, ensuring their unwavering support and satisfaction. This awe-inspiring aspect of EI can be the catalyst that propels your project management to new heights.

Developing Emotional Intelligence Skills

Developing your emotional intelligence is an ongoing process that involves continuous learning and practice. Here are some methods to enhance your EI skills:

One effective way to develop EI is through feedback mechanisms. Regular feedback from peers, mentors, or through self-assessment tools can provide valuable insights into how your emotions and actions affect others. This feedback is crucial for self-awareness and helps identify areas for improvement.

Self-reflection exercises also play a critical role in developing EI. Reflecting on your interactions and experiences can help you better understand your emotional responses and learn how to manage them more effectively. Techniques such as journaling or meditation can facilitate this self-reflection.

Formal training programs focused on emotional intelligence can also provide structured learning and practical exercises to enhance specific EI skills. These programs often cover emotional awareness, empathy, communication strategies, and stress management.

Case Studies on EI in Projects

Real-world case studies vividly illustrate the transformative impact of emotional intelligence in project management. Consider a scenario where a project failed due to poor team cohesion and communication breakdowns. The project manager, recognizing these issues, focused on building emotional intelligence within the team. The team's dynamic improved markedly through workshops and team-building exercises to enhance empathy and communication. This inspiring transformation restored project momentum and led to a more collaborative and innovative working environment.

Consider a real-life scenario in which a project manager, faced with significant changes in the project scope, used emotional intelligence to navigate the situation. Initially, the team was resistant to these changes. However, by employing empathy and effective communication, the manager was able to understand the team's concerns and motivate them toward the new objectives. This turned potential dissent into a unified effort toward project success, demonstrating emotional intelligence's practical application and benefits in project management.

These examples underscore the immense value of emotional intelligence in project management. It can lead to more effective leadership, smoother change management, and more robust team performance. By prioritizing the development and application of EI in your project management practices, you can enhance your projects' success and the well-being and growth of your team. Additionally, understanding and applying earned value management, which measures project performance and progress, can further boost your project management capabilities.

3.5 Negotiation Skills for Project Managers

Effective negotiation is not just a skill, but a cornerstone of successful project management. It plays a crucial role in establishing and maintaining the quality of relationships and terms with stakeholders and suppliers throughout a project's lifecycle. Mastering negotiation requires a blend of preparation, strategic communication, and tactical flexibility to navigate the complexities of project constraints and stakeholder expectations, underscoring its importance in project management.

Negotiation Techniques for Project Managers

The foundation of strong negotiation skills in project management lies in meticulous preparation. Before entering any negotiation, gathering as much information as possible about the other party's needs, priorities, and constraints is essential. This preparation involves understanding their business objectives, pressures, and what they value most in a partnership. Equally important is a clear understanding of your project's requirements, limits, and priorities. With this knowledge, you can develop a negotiation strategy to find common ground and mutual benefits rather than approaching the negotiation as a zero-sum game.

In terms of communication styles, effective negotiators are adept at adjusting their approach based on the audience and situation. This might mean being more assertive when dealing with a supplier accustomed to aggressive negotiation tactics or adopting a more collaborative and consultative style to win over a new stakeholder. The key is to remain respectful and professional, keeping the discussions focused on issues rather than allowing them to become personal.

Common negotiation tactics include anchoring, where you set the terms initially offered as a reference point for the negotiations; framing, which involves presenting information in a way that highlights the benefits in terms favorable to the other party; and making concessions strategically, where you give up something of lesser importance to gain something more valuable for the success of the project. These tactics and effective communication can significantly enhance your ability to negotiate favorable terms.

Negotiating Contracts and Terms with Suppliers and Stakeholders

When negotiating contracts and terms with suppliers and stakeholders, the goal is to achieve agreements that meet the project's needs while maintaining strong, cooperative relationships. Start by clearly outlining your project's requirements, budget constraints, and timelines. Use this as a framework to discuss contracts and terms. Communicating your needs openly and listening actively to the other party's concerns and limitations is essential. This two-way communication fosters a sense of partnership rather than opposition, emphasizing the value of these relationships in project success.

Negotiating contracts often involves discussing detailed and sometimes sensitive information. Ensuring confidentiality and demonstrating integrity in these discussions can help build trust—a key component of effective negotiations. Moreover, always be prepared to propose alternatives. For example, if budget limitations are a sticking point, suggest longer payment terms or a phased project rollout that might be more manageable within the financial constraints.

Maintaining good relationships ensures that all parties feel respected and that the agreements reached are perceived as fair. This might mean compromising on specific issues so that all parties feel like they have gained something of value. Remember, successful negotiations often lead to long-term collaborations, so the goodwill you foster during these discussions can benefit your projects in the future.

Handling Difficult Negotiations

Complex negotiations are inevitable in project management. Whether it's due to conflicting interests, budget constraints, or challenging personalities, it is crucial to have strategies to handle these situations. One practical approach is keeping the conversation focused on interests rather than positions. By understanding the underlying reasons for the other party's stance, you can often find solutions that satisfy the core interests of both sides.

When negotiations become particularly challenging, involving a neutral third party can be beneficial. This could be a mediator who can facilitate more productive dialogue and help both parties find a mutually acceptable solution. Another strategy is to take a break from negotiations to allow all parties to reassess their positions and the importance of various issues. This can help de-escalate tensions and provide fresh perspectives on how to move forward.

Role of Negotiation in Project Success

Mastering negotiation skills is not just about acquiring a skill, but about gaining a powerful tool that can empower project managers. It has a profound impact on everything from the project's budget and timeline to the quality of goods and services obtained and the strength of stakeholder relationships. Skilled negotiators can secure the resources, services, and cooperation essential for project success while managing costs and minimizing risks, giving them a strong sense of control and capability.

Real-life scenarios where negotiation skills have directly influenced project outcomes serve as compelling evidence of their significance. Consider a project manager negotiating the terms of a major software implementation. Their adept negotiation skills could secure additional support services at no extra cost, ensuring the project team can access expert help during the critical implementation phase. This keeps the project on track and strengthens the relationship with the supplier. Such successful negotiations can potentially lead to more favorable terms in future dealings, underscoring the practical value of negotiation skills in project management.

You can significantly enhance your effectiveness as a project manager by continuously refining your negotiation skills and applying them thoughtfully within the context of your projects. This leads to more successful project outcomes and builds your reputation as a competent and reliable leader capable of navigating complex negotiations to secure the best possible advantages for your projects.

3.6 Using Earned Value Management for Performance Tracking

Introduction to Earned Value Management (EVM)

In the fast-paced world of project management, maintaining a clear understanding of your project's health is paramount, and this is precisely where Earned Value Management (EVM) shines. A robust quantitative technique, EVM integrates scope, time, and cost data, providing a comprehensive snapshot of project performance and progress. The power of EVM lies in its ability to combine measurements of project performance related to cost and schedule, enabling you to assess your project's current status, forecast future performance, and identify potential issues before they escalate. This makes EVM an invaluable tool for project managers striving for efficient and successful delivery. Understanding the role of technology and innovation in project management is also crucial, as it is an emerging trend with the potential to enhance project efficacy and innovation.

At its core, EVM is built on three key data points: Planned Value (PV), which is the budgeted amount for the work scheduled to be completed by a specified date; Earned Value (EV), which is the budgeted amount for the work completed by that date; and Actual Cost (AC), which is the real cost incurred for the work completed. By comparing these values, EVM provides a clear and objective insight into project health beyond the simple budget versus actual analysis.

Setting Up an EVM System

The journey to implementing an EVM system begins with a solid foundation in planning. The first crucial step is to define the Work Breakdown Structure

(WBS), a hierarchical breakdown of the work to be executed by the project team. This structure breaks down the project into manageable sections, making it easier to assign budgets, schedule work, and measure progress. Once the WBS is in place, the next step is to set performance baselines, which are the agreed-upon plans that include scope, schedule, and cost baselines. These baselines serve as the benchmarks against which project performance is measured, providing a clear roadmap for successful project execution.

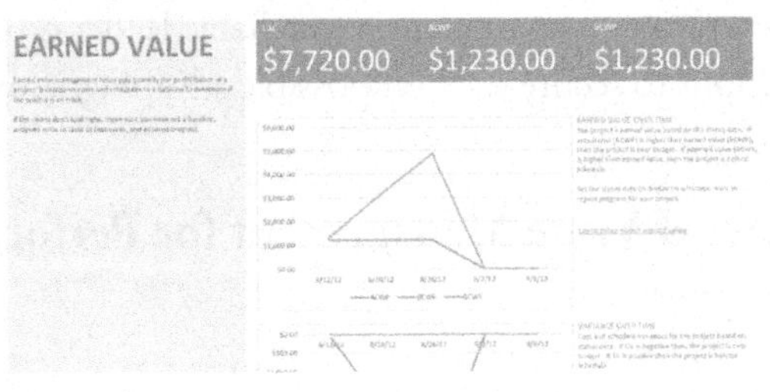

The selection of appropriate Key Performance Indicators (KPIs) is crucial. These metrics will help you track efficiency, productivity, and schedule adherence, among other factors. Common KPIs used in EVM include the Cost Performance Index (CPI) and Schedule Performance Index (SPI), which are ratios that measure the efficiency of the budget and schedule against the actual performance. Setting up these KPIs requires a deep understanding of the project's goals and the critical factors that lead to its success.

Analyzing Performance Data

Once your EVM system is in place, the next step is to analyze the performance data it generates regularly. This involves calculating cost variance (CV) and schedule variance (SV). CV is the difference between Earned Value and Actual Cost (CV = EV - AC). A positive CV indicates you are under budget, while a negative CV means you are over budget. Similarly, SV is the difference between Earned Value and Planned Value (SV = EV - PV). A positive SV indicates you are ahead of schedule, whereas a negative SV shows a delay.

Interpreting these metrics allows you to make informed decisions about the project. For instance, if you detect a consistent negative cost variance, it might be an indicator to reassess your budget allocation or review your spending to find areas where you can cut costs without compromising project quality. Similarly, a negative schedule variance would prompt a review of your project timeline and may lead to adjustments in work allocation or processes to get back on track.

Benefits and Challenges of EVM

EVM offers a host of benefits. It is an early warning system for off-track projects, enabling timely corrective actions. It enhances the accuracy of project forecasts in terms of cost and schedule and improves the project team's understanding of project performance. However, implementing EVM can be challenging. It requires rigorous up-front planning and baseline setting, which can be time-consuming. The accuracy of EVM data heavily relies on the accuracy of progress updates and cost tracking. It also requires a certain level of expertise to interpret EVM data correctly and make appropriate project decisions.

Despite the challenges, the strategic advantages of using EVM make it a worthwhile investment for any project manager. By understanding and leveraging EVM effectively, you can ensure that your projects are completed within budget and on time and meet or exceed stakeholder expectations. This strategic approach to project management can instill confidence in your decision-making and enhance your tracking and control mechanisms, making you feel more in control and strategic in your project management role.

As we wrap up this exploration into advanced project management strategies, remember that the tools and techniques discussed are theoretical concepts and practical solutions that can significantly enhance your project management capabilities. Each element elevates your ability to manage projects successfully, from optimizing resource allocation to mastering negotiation skills and implementing sophisticated tracking systems like EVM. As we move forward, let's carry these insights into our subsequent discussions, continuing to build on our foundation toward becoming adept project managers capable of handling complex challenges with confidence and skill.

Chapter 4

Technology and Innovation in Project Management

In the ever-evolving landscape of project management, the infusion of technology and innovation is not just a trend but a game-changer. It enhances project efficacy and aligns them with future advancements, offering a promising outlook for project managers. This chapter dives into the pivotal role that project management software plays in streamlining processes, fostering collaboration, and ensuring that projects meet their timeframes and budgetary constraints with higher precision. As you navigate through this digital augmentation, you'll discover how to select the right tools that resonate with the unique dynamics of your projects and teams, sparking optimism about the future of project management and making you feel excited about the potential of these new technologies.

4.1 Project Management Software: Choosing the Right One for Your Needs

Assessing Software Needs

Empower yourself as a project manager by carefully analyzing your project's scale, team distribution, and specific needs. This process allows you to take control of the software selection, ensuring that it aligns with your project's unique requirements. Start by evaluating the size of your project. Larger projects with multiple teams and stakeholders may require software with extensive features like advanced reporting capabilities and resource management tools. In contrast, smaller projects might benefit from more streamlined, user-friendly software with essential task management and communication features.

Consider your team's distribution. Are team members working remotely or based in the same office? Engage your team in the software selection process by discussing the importance of software that provides robust real-time collaboration features, such as cloud-based access and mobile compatibility. This is crucial for geographically dispersed teams to ensure everyone can update their progress and access project documents from anywhere.

Also, reflect on the specific features most critical to your project's success. Do you need software that integrates easily with other tools already used by your organization, such as ERP systems or financial software? Are real-time analytics and reporting capabilities essential to help you keep a pulse on the project's health? Make a list of must-have features to support your project's objectives and streamline operations.

Comparative Review of Popular Software

To illustrate, let's compare three widely used project management software platforms: Microsoft Project, JIRA, and Asana. Microsoft Project is renowned for its comprehensive project planning features, resource allocations, and detailed reporting. It's particularly well-suited for large-scale projects that require meticulous scheduling and resource management. However, its complexity and robustness might be overwhelming for smaller projects or those new to project management software.

Initially designed for software development projects, JIRA excels in task management, bug tracking, and iterative work processes like Agile and Scrum methodologies. Its ability to customize workflows and boards makes it a favorite among teams that value flexibility and adaptability in managing tasks.

Asana, on the other hand, is known for its user-friendly interface and simplicity. It provides essential tools for task management, team collaboration, and deadline tracking, making it ideal for smaller teams or projects that don't require complex project management functionalities but prioritize ease of use and quick setup.

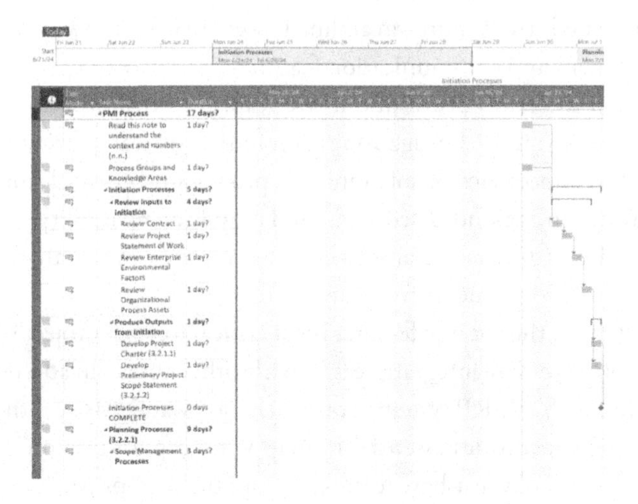

Cost-Benefit Analysis

Conducting a cost-benefit analysis is crucial in ensuring that the chosen software provides value for money while meeting your project's needs. Start by estimating the cost associated with each software option, including subscription fees, training costs, and additional costs for support or extra features. Then, assess each software's benefits to your project, such as time saved through efficient task management, improved communication among team members, or reduced overheads from automating specific processes.

Consider both direct and indirect benefits. For instance, software with a higher upfront cost but offering extensive integration capabilities could reduce long-term expenses associated with using multiple disjointed tools. Similarly, platforms with excellent support services might prevent costly downtime and ensure smoother operation.

Implementation Tips

Successfully implementing new project management software is not just about technical setup; it's about ensuring that all users are equipped to use the latest tools effectively. This starts with comprehensive training that covers all the software's features and functionalities. By tailoring training sessions based on the roles and responsibilities of different team members, you can ensure that each user understands how to use the new tools effectively, maximizing the software's potential and minimizing user frustration.

Integration into existing workflows should be handled with care to avoid disruption. Start with a pilot program, selecting a small team or project to test the new software. This approach allows you to gather insights and make necessary adjustments before rolling it out to the entire organization.

Continuously gather feedback during and after implementation to gauge the software's effectiveness and user satisfaction. Be prepared to make iterative improvements to ensure the software fully meets the needs of your projects and teams.

By thoughtfully assessing needs, comparing options, conducting a thorough cost-benefit analysis, and carefully planning the implementation, you can ensure that the project management software you choose fits your immediate needs, scales with your projects, and enhances overall productivity and collaboration.

4.2 The Role of AI and Machine Learning

Learning in Project Management

Artificial Intelligence (AI) and Machine Learning (ML) are reshaping the landscape of project management by offering tools that not only automate routine tasks but also enhance decision-making and risk management. In essence, AI refers to the capability of machines to perform tasks that typically require human intelligence, such as recognizing patterns, learning from experience, and making decisions. Machine Learning, a subset of AI, involves algorithms that enable computers to learn from and make predictions based on data. Integrating these technologies into project management can significantly impact how projects are planned, executed, and monitored, leading to more efficient and effective management processes.

One of the most compelling applications of AI in project management is in the domain of risk management. Traditional risk management methods involve manually identifying and tracking potential risks based on historical data and team input. However, AI can revolutionize this process by automatically analyzing vast amounts of data to identify risk patterns and predict potential issues before they impact the project. For example, AI algorithms can analyze data from previous projects to identify factors that led to budget overruns or delays and alert managers about similar risks in current projects. This proactive approach to risk management saves time and resources and enhances the project's likelihood of success by allowing timely mitigation strategies.

Additionally, Machine Learning algorithms can significantly enhance decision-making in project management. These algorithms analyze historical project data and ongoing performance metrics to provide insights that inform better decision-making. For instance, ML can help project managers predict the optimal allocation of resources for various tasks, forecast project timelines based on progress data, and suggest the best communication strategies to keep stakeholders engaged. By providing data-driven insights, ML helps project managers make more informed, objective, and timely decisions, thereby increasing the efficiency and effectiveness of project management practices.

To illustrate the real-world impact of AI and ML in project management, consider the example of a large construction company that implemented ML algorithms to optimize its resource allocation and workflow scheduling. The ML model analyzed data from several past projects, including resource use, project timelines, and the outcomes of different scheduling strategies. By learning from this data, the model provided resource allocation and task scheduling recommendations that minimized downtime and avoided conflicts between simultaneous tasks. As a result, the company saw a 15% decrease in project completion time and a 20% reduction in labor costs, demonstrating the substantial benefits of integrating ML into project management.

Another case involves a software development firm that used AI to enhance its risk management processes. The AI system was trained to analyze code commits, project logs, and developer comments to identify patterns that previously led to security flaws or bugs. By flagging these risks early in the development process, the team could address potential issues long before they become costly or damaging. This proactive approach not only improved the quality of the software but also reduced the time spent on debugging and revisions, highlighting the transformative potential of AI in project management risk assessment.

These illustrative scenarios vividly demonstrate the revolutionary role that AI and ML are playing in the realm of project management. By streamlining complex analytical tasks, delivering insights with unprecedented speed, and significantly improving the quality of decision-making, these cutting-edge technologies empower project managers to steer more intricate projects toward successful completion with enhanced efficiency and certainty. Automating laborious analyses and providing real-time, actionable intelligence allows project managers to anticipate hurdles and strategize effectively, ensuring projects remain on track and exceed expectations. As the sophistication of AI and ML technologies advances, their incorporation into the arsenal of project management tools and methodologies is expected to deepen, further elevating the proficiency and impact of project managers in diverse sectors. This evolution signifies a promising horizon where the integration of AI and ML into project management becomes more widespread and fundamentally transforms the landscape of project execution and oversight, offering remarkable benefits in productivity, cost-efficiency, and project outcome predictability.

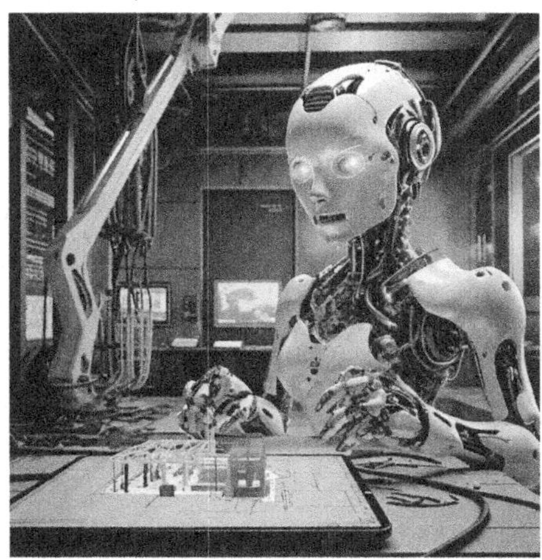

AI in Project Management

4.3 Integrating New Technologies into Your Project Management Practice

As the landscape of project management continually evolves, staying ahead involves adopting new technologies and ensuring they seamlessly integrate into your existing practices. Evaluating new technologies should begin with clearly understanding how they align with your project management goals and the specific challenges they aim to address. Key considerations include the technology's scalability, compatibility with existing tools, and potential to enhance project efficiency and effectiveness. For instance, a new scheduling tool should offer superior features and integrate smoothly with your current project management suite to improve your team's productivity.

When assessing a new technology, define the most important criteria for your team's needs and project outcomes. These might include the speed of implementation, ease of use, required training, customer support, and cost. Engage with stakeholders, including your project team, IT department, and financial officers, to ensure all perspectives are considered. This comprehensive approach ensures that the selected technology not only meets the immediate needs of your project team but also aligns with the broader organizational goals and IT strategy.

Once a technology has been selected, developing a well-structured integration strategy is crucial. Begin by setting clear objectives for your goal with the new technology. These objectives should be specific, measurable, achievable, relevant, and time-bound (SMART). Next, create a phased rollout plan for gradual technology implementation across projects. This might involve starting with a pilot project as a test case to evaluate the technology's impact and identify any potential issues before broader implementation.

Training and preparing your project team for the new technology is essential for its adoption and utilization. Invest in comprehensive training programs that cover the technical aspects of the new tool and address changes in processes and workflows that its implementation will entail. Identifying or hiring tech-savvy champions who can help their colleagues adapt to the new system is also beneficial. These champions can provide on-the-ground support and feedback, facilitating a smoother transition and encouraging widespread acceptance of the latest technology.

Measuring the impact of technology integration on your project management practices involves tracking specific metrics that reflect the technology's contribution to project success. These metrics might include the time saved in project scheduling, data accuracy improvements, or team members' communication speed. Set up a system to collect and analyze this data regularly, allowing you to quantify the benefits and identify areas for further improvement.

Additionally, the qualitative benefits of the new technology, such as improved team morale and increased stakeholder satisfaction, should be considered. Regular surveys and feedback sessions can help gauge these less tangible outcomes, providing a holistic view of the technology's impact. By continuously monitoring and adjusting your approach based on these insights, you can ensure that integrating new technologies into your project management practice drives substantial value and supports your team in achieving project excellence.

4.4 Virtual and Remote Project Management: Tools and Tips

Navigating the intricacies of virtual and remote project management introduces a distinct set of challenges that can significantly impact the dynamics and success of your projects. One of the primary hurdles you might encounter is overcoming communication barriers. Unlike traditional settings where face-to-face interactions facilitate quick exchanges and clarity, remote environments can often lead to miscommunications or delayed information flow. Additionally, the lack of physical supervision in remote settings can sometimes decrease visibility over team members' progress and daily activities. These challenges can lead to project delays, reduced productivity, or disengaged teams if not appropriately managed.

Adopting specific tools and software designed for remote project management is essential to manage these challenges effectively. Platforms like Zoom or Microsoft Teams have become staples for virtual meetings, offering features replicating an in-office meeting environment with video calls, screen sharing, and real-time chat functionalities. Cloud-based software such as Smartsheet or Basecamp can be invaluable for project tracking and management. These tools allow you to create, assign, and track tasks across your team, no matter where members are located. They also provide integrative features allowing seamless connections with other tools, enhancing data flow and accessibility.

Collaboration platforms like Slack or Microsoft Teams facilitate continuous communication, which is crucial for maintaining project momentum and team cohesion. These platforms support the creation of specific channels for different projects or topics, making it easy to keep discussions organized and relevant. They also integrate with other tools, ensuring that important documents and information are readily accessible to all team members. Implementing these tools can significantly reduce the risks associated with remote project management by enhancing communication and ensuring everyone stays on the same page.

Best Practices for Managing Remote Projects

Several best practices can be adopted to further refine your approach to virtual project management and ensure effectiveness and efficiency. Establishing clear communication protocols is fundamental. This involves defining specific channels for daily communications, urgent messages, and different types of updates. It also means setting expectations about responsiveness and available hours, especially if team members are in different time zones.

Regular check-ins are another critical practice; as a team leader, you play a pivotal role in conducting them. Whether daily or weekly, depending on the project's pace and complexity, these sessions are designed to discuss progress and any challenges or roadblocks the team members might be facing. This consistent engagement helps quickly address issues before they escalate and promotes a sense of inclusion and support for remote team members, making them feel valued and integral to their operations.

Virtual leadership is pivotal in steering projects toward success, especially in remote work. It's not just about managing tasks; it's about understanding and addressing remote workers' unique challenges, such as isolation or work-life balance issues. As a virtual leader, your accessibility, encouragement, and support are crucial in fostering a positive team culture that values each member's contribution.

Engagement Strategies to Keep Remote Teams Motivated

Keeping a remote team engaged and motivated requires intentional strategies beyond traditional project management techniques. One practical approach is to foster a strong team culture, even if it's virtual. For example, at Franklin Publications, we celebrate milestones and successes in team meetings, create virtual hangout sessions that are not about work, and encourage informal interactions among team members. These activities help build relationships and community, often needing to be added in remote settings.

Additionally, recognizing and rewarding good work can profoundly impact morale. This might include shout-outs during meetings, written acknowledgments in team channels, or even small bonuses for project milestones. Such recognition boosts the individual's morale and sets a positive tone across the team, showing that hard work is noticed and valued.

Providing opportunities for professional growth is crucial. This could involve virtual training sessions, course access, or assigning responsibilities that stretch their capabilities. When team members feel they are growing professionally and have paths for advancement, even in a remote setting, their engagement and satisfaction levels are likely to be higher. This not only benefits the individual team members but also the project as a whole. For instance, when team members are more engaged and satisfied, they're more likely to contribute innovative ideas and solutions, improving project outcomes.

By integrating these tools and adopting these best practices and engagement strategies, you can effectively manage the unique demands of virtual and remote project management. However, it's important to note that there may be challenges along the way. For instance, some team members may struggle with transitioning to virtual hangout sessions or find balancing work and personal life in a remote setting challenging. To overcome these challenges, it's crucial to communicate openly with your team, listen to their concerns, and adjust your strategies accordingly. This ensures your projects stay on track and maintains high team morale and productivity levels, crucial for long-term success in an increasingly digital work environment.

4.5 Cybersecurity Considerations for Project Managers

In today's digital landscape, where technology seamlessly integrates into project management, understanding cybersecurity fundamentals is beneficial and empowering. Cybersecurity in project management is about more than just protecting data and maintaining stakeholder trust. It's about ensuring that you, as a project manager, are not at the mercy of the technological tools and methods employed, but rather, that you are in control, with the knowledge and skills to prevent them from becoming the weakest link in your project management chain.

The first step in embedding cybersecurity into your project management practice is to grasp the basic concepts for building a secure infrastructure. Cybersecurity involves protecting systems, networks, and programs from digital attacks. These attacks usually aim to access, alter, or destroy sensitive information, extort money from users, or interrupt normal business processes. For instance, a common type of attack is a 'phishing attack, 'where fraudulent communication comes from a reputable source to steal sensitive data. Implementing effective cybersecurity measures is particularly challenging today because there are more devices than people, and attackers are becoming more innovative.

Recognizing the potential cyber threats that could impact your project is a critical skill. Common threats include data breaches, where sensitive information is accessed without authorization; phishing attacks, where fraudulent communication appears to come from a reputable source to steal sensitive data; and ransomware, a type of malicious software designed to block access to a computer system until a sum of money is paid. Identifying these threats involves staying informed about the latest cybersecurity trends and understanding how these threats could be applied to your project. Regularly updating your knowledge base and maintaining a proactive approach to potential vulnerabilities within your project's digital tools can significantly mitigate these risks.

Implementing practical cybersecurity measures involves several steps but starts with the fundamentals: ensuring secure communication channels within your project team. Here's a step-by-step guide on how to do this: **Step 1:** Identify the communication channels used by your team. **Step 2:** Research and select encrypted messaging apps and secure email services that protect against interception and unauthorized access to sensitive project communications. **Step 3:** Train your team on how to use these tools effectively. Data encryption is another critical step, ensuring all project data stored and transmitted is encoded and can only be accessed or decrypted by individuals with the correct encryption keys. Regular security audits and assessments should also be part of your routine, examining your project's infrastructure and processes to identify vulnerabilities that cyber attackers could exploit.

Developing an effective incident response plan is your safety net in a security breach. This plan should outline the procedures to follow when a cybersecurity incident occurs, detailing how to contain the breach, assess and manage the damage, and communicate with stakeholders during and after the incident. The plan should also include the steps to restore any disrupted services and processes as quickly as possible to minimize impact on the project. Regular training sessions with your project team on recognizing and responding to security threats are essential, ensuring everyone knows their role in the incident response plan. This proactive preparation can be the difference between a minor disruption and a major crisis in your project management.

By integrating these cybersecurity practices into your project management framework, you enhance not only the security of your projects but also the confidence of your clients and stakeholders in your ability to manage and protect their interests. In an era where digital threats are evolving rapidly, equipping yourself with the knowledge and tools to mitigate these risks is crucial for the modern project manager.

4.6 Case Study: Implementing Blockchain for Enhanced Project Transparency

Introduction to Blockchain Technology

Blockchain technology, often associated with cryptocurrencies like Bitcoin, is a decentralized digital ledger that records transactions across multiple computers so that the registered transactions cannot be altered retroactively. This technology offers robust transparency and security, making it highly relevant to project management, particularly in aspects that require unalterable records and traceability. However, it's essential to be aware of the potential costs. For instance, implementing a blockchain network can be expensive, especially starting from scratch. To manage these costs, it's crucial to budget effectively and consider the long-term benefits of implementing blockchain technology.

Blockchain for Transparency and Accountability

In project management, the potential impact of blockchain technology is not just significant- it's transformative. Every transaction or task update recorded in a visible and unchangeable way can significantly reduce the risks of discrepancies and fraud. Imagine the possibilities in large construction projects involving numerous subcontractors and suppliers, where blockchain can provide a transparent record of materials supplied, work completed, and payments made, ensuring all parties have access to the same information and reducing conflicts or misunderstandings.

Blockchain's capability to automate specific processes through smart contracts — self-executing contracts with the terms of the agreement directly written into code — can further enhance efficiency and accountability. These contracts automatically enforce obligations and trigger actions like payments or notifications when certain conditions are met, streamlining project management tasks and reducing the administrative burden on project teams.

Real-World Implementation

Let's dig into the transformative power of blockchain technology, as demonstrated by a multinational development firm. They implemented blockchain to enhance transparency and efficiency in a large-scale infrastructure project spanning several countries. This project involved multiple stakeholders, each requiring access to timely and accurate project data. The firm's use of blockchain revolutionized the project management process, allowing for real-time updates, secure data sharing, and automated contract enforcement.

The firm utilized a blockchain platform to manage all project-related transactions, from procurement to subcontractor payments. Each transaction was recorded as a block on the blockchain, linked to previous transactions, and visible to all parties. This provided a transparent view of the project's financial flows and significantly reduced payment delays and disputes over services rendered.

However, the implementation faced several challenges. The primary issue was the resistance from stakeholders unfamiliar with the technology and hesitant about its benefits. To address this, the firm conducted extensive training sessions and workshops to educate stakeholders about how blockchain works and how it would simplify their workflow and payment processes.

The outcome was overwhelmingly positive. The blockchain implementation led to a 30% reduction in administrative costs and a significant decrease in payment disputes. The project was completed ahead of schedule, and the transparency provided by blockchain enhanced trust among all parties, setting a new standard for managing large-scale international projects.

Assessing the Benefits and Limitations

While implementing blockchain in project management can bring significant benefits such as enhanced transparency, increased security, and improved efficiency, it's equally important to understand its limitations and challenges. Blockchain technology necessitates a substantial initial investment in setup and training, and a robust IT infrastructure and ongoing maintenance are prerequisites for managing the blockchain system. By providing a balanced view of the benefits and limitations, you can assist readers in making informed decisions about whether blockchain aligns with their project management needs.

Additionally, the success of blockchain implementation hinges on the active participation and buy-in from all stakeholders. As the case study illustrates, stakeholder resistance can pose a significant hurdle. Therefore, a meticulous analysis of the project's context, the readiness of the involved parties, and the projected return on investment is vital before embarking on implementing blockchain technology in project management.

In conclusion, blockchain offers exciting possibilities for enhancing transparency and accountability in project management. While it comes with challenges and limitations, the potential to streamline processes, reduce fraud, and build trust makes it a worthy consideration for projects requiring high transparency and involving multiple stakeholders. As we continue exploring innovative project management technologies, understanding and leveraging tools like blockchain will be crucial in shaping the future of effective and efficient project management practices.

The Project Management Blueprint Review Request

Title: Make a Difference with Your Review

Subtitle: Unlock the Power of Helping Other Project Managers

"The art of project management is making complexity look simple." - Scott Berkun

Here's a question for you...

Would you help someone you've never met, even if no one knew you did it?

Who is this person? They're a lot like you were not too long ago. They want to learn about project management but don't know where to start. They need help but aren't sure who to ask.

We want to make project management easy for everyone to understand. Everything we do is to reach that goal. But we can only do it if we reach... well... everyone who needs help.

This is where you come in. Most people really do judge a book by its cover (and its reviews).

Your gift doesn't cost any money and takes less than a minute, but it could change someone's life forever. Your review could help...

...one more small business finish an important project.

...one more team leader organize their work better.

...one more employee get a promotion.

...one more company make their customers happy.

...one more dream come true.

To get that warm, fuzzy feeling of helping someone takes less than 60 seconds, and all you have to do is...

leave a review.

Just scan this QR code to leave your review:

Or click on the link(Ebook Readers)

[https://amazon.com/review/create-review?&asin=B0DBQYH5K4]

If you feel good about helping a project manager you've never met, you're our kind of person. Welcome to the club. You're one of us!

I'm so excited to help you become a great project manager faster and easier than you ever thought possible. You'll love the tips and tricks I'm about to share in the next chapters.

Thank you from the bottom of my heart. Now, let's get back to learning about project management!

- Your biggest fan, Franklin Publications

Chapter 5

Leadership and Team Dynamics

In the intricate choreography of project management, the team sets the rhythm, and as the project manager, you are the conductor. It's your responsibility to ensure that every team member moves in harmony, contributing their unique skills to the collective performance. Understanding how to build and lead effective project teams is crucial for the project's success and for fostering an environment where everyone can thrive and grow. Here, we'll delve into how to assemble your ensemble—selecting the right performers, setting the stage for collaboration, and directing a masterpiece of teamwork and productivity.

5.1 Building and Leading Effective Project Teams

Team Formation Principles

As a project manager, your role is pivotal in guiding your team through the four stages of team formation: forming, storming, norming, and performing. Each stage represents a step in the journey toward becoming a cohesive unit. Your task is to provide insight and tact, ensuring the team progresses smoothly through these phases. This underscores the importance of your leadership in shaping the team dynamics and, ultimately, the project's success.

In the **forming** stage, team members are introduced. They start learning about the project and their responsibilities, forming initial impressions. This is your opportunity to set a positive tone and clear expectations. Facilitating introductions and encouraging open communication is crucial so team members feel comfortable with each other and their roles.

As the team moves into the **storming** stage, the initial politeness often gives way to a jostling for positions; personal and working style differences can surface, leading to conflict. This stage tests your skills in diplomacy and conflict resolution. Your role as a project manager is not just about remaining impartial and facilitating healthy, constructive discussions on disagreements. It's also about encouraging team members to express their concerns openly, ensuring they feel heard and validated, which can prevent resentments from festering. Your leadership in this stage is crucial in shaping the team dynamics and, ultimately, the project's success.

Following the storming phase, the team enters **norming**, where members resolve their differences, appreciate colleagues' strengths, and respect their authority as leaders. Cohesion begins as everyone works more effectively and adheres to established norms and practices. During this phase, reinforce positive behavior by recognizing achievements and continued collaboration.

Finally, in the **performing** stage, the team reaches its full potential. Members are confident and motivated and work collaboratively towards the project's goals. Your role here shifts more towards maintaining momentum and facilitating continuous improvement. This is the stage where the team's collective skills and the synergy you've helped foster shine brightest.

Navigating these stages isn't always linear and may require you to adapt your leadership approach based on the team's dynamics and the project's needs.

Team Building

Role Assignments and Responsibilities

Assigning the right roles to the right people is crucial in leveraging your team's diverse strengths. A useful tool here is the RACI matrix, which stands for Responsible, Accountable, Consulted, and Informed. It's a straightforward tool that helps clarify roles and responsibilities in a project, ensuring everyone knows what they are expected to do and whom they need to inform or consult on different tasks.

When you assign roles, align them with team members' strengths and project needs. Some might excel in data analysis but struggle with public presentations, and others might thrive in client interactions but not in detailed report writing. By aligning roles with individual strengths, you enhance productivity and increase job satisfaction among team members.

Leadership Styles and Team Dynamics

Your leadership style plays a pivotal role in shaping the team dynamics. Whether you adopt an **authoritative** style, making quick decisions and providing clear direction, or a **democratic** style, involving team members in decision-making, each style impacts the team differently and can be effective depending on the situation and team composition.

For instance, authoritative leadership might be effective when quick, decisive action is needed in a crisis. In contrast, democratic leadership can be more appropriate when the project scope is ambiguous and you need various creative solutions. There's also **laissez-faire** leadership, where team members are given autonomy over how to complete their tasks. This style can foster innovation and independent problem-solving but requires you to trust your team's abilities and be ready to step in if guidance is needed.

Creating a Collaborative Environment

Fostering a collaborative team environment is essential for project success. Encourage open communication by establishing regular check-ins and creating a safe space for sharing ideas and concerns. Be transparent about project challenges and successes to promote trust and mutual respect among team members.

Emphasize the significance of team-building activities, even those unrelated to work, in fostering collaboration. These activities can break down barriers, foster connections, and enhance teamwork among project members. Additionally, collaborative tools that empower you to share information and provide progress updates can ensure everyone is aligned and contributing effectively to the project's goals.

By carefully managing these aspects of team dynamics and leadership, you create a team that can achieve exceptional results and a work environment that is dynamic, respectful, and geared toward continuous growth and improvement. As you lead your team through the complexities of project execution, remember that your role as a facilitator of their talents and ideas is just as important as your expertise in project management. This approach not only drives project success but also builds a team that is robust, resilient, and ready for any challenges ahead.

5.2 Conflict Resolution Strategies for Project Managers

In the dynamic arena of project management, conflicts are not just inevitable; they are a natural part of the interactions between diverse individuals working under pressure to meet common goals. Recognizing the familiar sources of these conflicts can drastically improve your ability to manage them effectively. Conflicts often arise from resource allocation, where team members feel they need more resources to achieve their tasks, or from tight deadlines that may seem unattainable, adding stress and friction among team members. Personal differences, rooted in diverse personalities, working styles, or values, also frequently lead to conflicts. Identifying these conflict sources early is crucial. It involves maintaining a close connection with your team, attentively listening to their feedback, and observing team interactions for any signs of tension or dissatisfaction.

Once you've identified potential or existing conflicts, employing the proper conflict resolution techniques can make a significant difference. Techniques such as mediation, arbitration, and negotiation are invaluable tools in your management arsenal. Mediation involves you, the project manager, acting as a neutral party to help those involved in the conflict find a mutually satisfactory solution. It's particularly effective when the conflict is rooted in misunderstandings or communication issues, as it allows each party to express their perspective in a controlled, respectful environment.

Arbitration, on the other hand, may be necessary when a conflict is more severe or when mediation fails. This technique involves a more authoritative approach where you resolve the conflict after hearing each party's side. While this can effectively resolve disputes, it may only sometimes be the best option for team harmony if parties feel their voices were adequately considered.

Negotiation is another critical technique beneficial when conflicts arise from resource allocation or deadlines. It involves direct dialogue between conflicting parties to reach a compromise that acknowledges the needs and limitations of everyone involved. Effective negotiation requires a deep understanding of each team member's priorities and the flexibility to explore alternative solutions that satisfy the project's requirements without compromising team cohesion.

Preventive measures are equally necessary in managing conflicts. Setting clear expectations from the start of the project, maintaining regular communication, and creating a platform for feedback can significantly reduce misunderstandings and grievances. Clear expectations ensure that all team members understand their roles, responsibilities, and the project's goals, reducing ambiguity that could lead to conflicts. Regular communication, through meetings or progress updates, keeps everyone aligned and informed, making it easier to address issues before they escalate into disputes. Additionally, providing a platform for feedback allows team members to express concerns or suggestions early, which can be addressed promptly to prevent potential conflicts.

To illustrate these strategies, consider a case where a project manager faced a significant conflict over resource allocation in a software development project. The development team felt the testing team needed more resources, leading to tension and reduced productivity. The project manager identified the conflict early through regular team updates and decided to mediate. By facilitating discussions, the manager helped both teams express their concerns and understand the resource distribution strategy. Realizing that the issue stemmed from a need for more transparency, the project manager negotiated a new agreement where resource allocation was more clearly communicated and aligned with project phases, satisfying both teams.

In another instance, a project manager dealt with a conflict arising from clashing personalities in a marketing project. Two team leaders disagreed on the project's creative direction, each staunchly advocating for their vision. The manager used arbitration, deciding on a direction that best aligned with the client's needs and project goals. To ensure continued collaboration, the manager also introduced regular collaborative workshops to foster a better understanding and respect for each other's contributions, significantly improving team dynamics.

These examples highlight the importance of resolving conflicts and taking proactive steps to prevent them. By fostering an environment of open communication, respect, and mutual understanding, you can minimize conflicts and their impact on your projects, leading to more successful outcomes and a more harmonious team environment.

Conflict Resolution

5.3 Motivating Your Team: Techniques and Theories

Understanding what drives your team members is pivotal in unleashing their full potential and achieving project success. Motivation in project management can be complex, as it involves aligning individual aspirations with the project's objectives. To navigate this, let's delve into some foundational theories of motivation that can provide a structured approach to understanding and influencing team behavior.

Maslow's hierarchy of needs, a staple in psychological studies, suggests that people are motivated by a hierarchy of needs: physiological, safety, love/belonging, esteem, and self-actualization. Applying this to project management, you can see that ensuring basic needs are met (like job security and working conditions) sets the stage for higher-level motivations such as recognition (esteem) and opportunities for personal growth (self-actualization).

Herzberg's two-factor theory, on the other hand, divides factors into 'hygiene' (factors that prevent dissatisfaction but don't necessarily motivate) and 'motivators' (factors that truly propel people to perform better). For instance, ensuring that workload management is fair can prevent dissatisfaction (hygiene), whereas recognizing someone's contribution in a team meeting could motivate morale and productivity.

Understanding McClelland's Theory of Needs, which categorizes motivation into achievement, affiliation, and power, can empower you as a project manager. Recognizing that different needs may drive each team member, you can tailor your motivational strategies more effectively. Some might strive for success and tangible achievements, others seek solid interpersonal relationships, and some might be motivated by influence and leadership opportunities. This understanding gives you the power to align your strategies with the intrinsic motivations of your team members, fostering a more productive and engaged team.

Tailoring Motivation Strategies

Implementing effective motivation strategies requires a nuanced approach considering the project context and the team members' differences. Start by setting clear, achievable goals directly linked to the project's outcomes. This gives team members a sense of purpose and a clear understanding of how their contributions impact the project. Setting incremental milestones with clear rewards can particularly motivate those driven by achievement.

Intrinsic motivators like personal growth and accomplishment are often more sustainable drivers than extrinsic motivators like bonuses or gifts. Encourage individual development by offering opportunities for skills training or leadership roles. For those motivated by affiliation, fostering a team-oriented environment where collaboration and support are emphasized can increase motivation significantly.

Recognition and Rewards Systems

Designing a recognition and reward system that aligns with the project goals and team values is crucial. This system should acknowledge achievements and reinforce the behaviors that lead to those achievements. Simple gestures such as thanking a team member during a team call for their hard work or highlighting a team's effort in a project newsletter can have profound effects. Consider more formal rewards such as bonuses, certificates, or public acknowledgment in company-wide meetings for more significant achievements.

Ensure that the criteria for recognition are transparent and consistently applied. This fairness is crucial in maintaining high morale and trust within the team. Tailor rewards are meaningful to the recipient; for instance, some might appreciate public recognition, while others prefer private acknowledgment or tangible rewards.

Maintaining High Morale

As a project manager, your role in maintaining morale is crucial, especially during long-term projects or challenging phases. Your responsibility goes beyond project updates; it's about gauging your team's well-being and job satisfaction. These check-ins, led by you, can provide insights into potential issues before they become demotivating, making your team feel valued and integral to the project's success. Recognizing your role and responsibility can make you feel more empowered and capable of leading your team effectively.

As a project manager, you can encourage a work culture that promotes work-life balance. Recognize that overworked team members are more likely to experience burnout, drastically affecting morale and productivity. Promoting a balanced approach demonstrates that you value their well-being, making them feel cared for and respected, increasing loyalty and motivation.

Celebrating small wins and milestones is a formality but a powerful tool for keeping the team's spirits high throughout the project. When done sincerely and inclusively, these celebrations remind everyone of the progress, even when the end goal might seem far away. They also serve as recognition of your team's efforts, making them feel motivated and appreciated. This emphasis on celebration can make your team feel more motivated and valued, leading to increased loyalty and productivity.

By understanding the underlying theories of motivation, you can better tailor your strategies to meet your team's diverse needs. This enhances individual performance and fosters a dynamic, supportive, and conducive project environment conducive to achieving outstanding results. As you continue to guide your team through the complexities of your projects, remember that a motivated team is not just about enhanced productivity but also about creating a work atmosphere that is vibrant, engaging, and profoundly fulfilling for everyone involved.

5.4 Stakeholder Management: Strategies for Engagement and Communication

In the complex world of project management, the art of stakeholder management is like conducting an orchestra. Like a musician, each stakeholder has a unique role in the symphony of your project's success. Identifying these key players and engaging with them effectively can significantly shape your project's path. The first crucial step is identifying who has a stake in your project's outcome. Stakeholders include internal teams, upper management, external clients, suppliers, and regulatory bodies. Each stakeholder has a different level of influence and interest, which can impact your project in various ways. To navigate this landscape effectively, use a stakeholder analysis matrix. This tool helps you visualize the power and interest of each stakeholder, enabling you to prioritize your engagement efforts. Stakeholders with high energy and interest require your utmost attention and frequent communication, as their input and satisfaction are vital to your project's progression.

Once you've identified and prioritized your stakeholders, developing a tailored engagement plan is your next pivotal step. This plan should outline specific strategies for how and when you will communicate with each stakeholder group. Tailoring your approach is crucial; for instance, technical details critical to your engineering team might be overwhelming for your marketing department or external partners. Similarly, regular, detailed updates might be appreciated by project sponsors but could be unnecessary for other groups, such as regulatory agencies, who may only require milestone-based reporting. The communication's clarity, timeliness, and relevance characterize effective stakeholder engagement. Tools such as personalized emails, newsletters, and regular briefings can be instrumental in keeping stakeholders informed and involved.

Managing stakeholder expectations is one of the most challenging aspects of stakeholder engagement. Clear communication about project goals, timelines, and potential risks is crucial in setting realistic expectations. It's also important to remain flexible and responsive to stakeholder concerns. Regularly seeking their feedback and showing readiness to adapt your strategies can help manage their expectations throughout the project lifecycle. For instance, if a project delay seems inevitable, proactively communicate this with a rationale and an updated timeline rather than allowing stakeholders to discover this independently. Such transparency maintains trust and helps manage their reactions and the overall impact on the project.

Building and maintaining solid relationships with stakeholders is fundamentally about fostering trust and ensuring transparency. Establish yourself as a reliable point of contact by consistently communicating and following through on commitments. Face-to-face interactions, though only sometimes feasible, can significantly enhance relationship building. They allow for more personal interaction and the opportunity to address complex issues more effectively. In situations where in-person meetings are not possible, video conferencing can be a viable alternative, providing a more personal touch than emails or phone calls. Remember, the goal is to make stakeholders feel valued and respected throughout the project process, reinforcing their importance to the project's success and fostering a cooperative rather than confrontational relationship. By maintaining transparency, you instill confidence in your stakeholders and your project management skills.

By mastering these aspects of stakeholder management, you enhance the prospects of your project's success and build a robust support network that can be invaluable for future projects. This proactive approach to managing interactions and communications with stakeholders ensures that challenges are navigated smoothly and that the project remains aligned with the needs and expectations of those it aims to serve.

5.5 Inclusive Leadership: Managing Diverse Teams

Embracing diversity within project teams is not just a matter of inclusivity; it's a strategic advantage. Diverse teams bring a wealth of perspectives, experiences, and skills, which are invaluable in tackling the complex challenges of modern projects. Each unique background opens up a broader range of ideas, fostering innovation and enhancing problem-solving capabilities. For instance, a team with members from different cultural backgrounds can provide insights into global markets, which is crucial for international projects. Similarly, diversity in skills and professional backgrounds ensures a thorough and well-rounded approach to project execution. By embracing diversity, you inspire your team to reach new heights of innovation and problem-solving.

However, leading a diverse team also presents unique challenges that require thoughtful strategies to manage effectively. Differences in cultural norms, communication styles, and work ethics can lead to misunderstandings and conflicts if not proactively addressed. As a project manager, it's critical to develop cultural competence, which involves understanding, respecting, and appropriately responding to the diversity present in your team. This starts with a commitment to learning about the cultural backgrounds of your team members and acknowledging the validity of their perspectives and experiences. It also involves being aware of your cultural biases and how they might influence your interactions and decisions.

Developing cultural competence extends beyond individual interactions and should be integrated into the project management processes and practices. This can involve implementing policies that respect cultural differences, such as flexible working hours to accommodate different time zones or religious practices or modifying communication practices to ensure that all team members can participate effectively, regardless of their linguistic or cultural backgrounds. For example, you might provide materials in multiple languages or use visual aids and clear, simple language to ensure that non-native speakers are not disadvantaged.

Inclusive communication practices are pivotal in managing diverse teams effectively. Ensuring all team members feel valued and heard is essential for maintaining morale and engagement. This involves actively soliciting input from all team members, particularly those less inclined to speak up in group settings. Techniques such as round-robin feedback sessions, where each team member is given the floor to express their thoughts, or anonymous suggestion boxes can help gather input from everyone. When discussing project details, use inclusive language that does not assume a specific cultural perspective. Be clear and precise to avoid misinterpretations leading to errors or resentment.

Inclusive leadership means being responsive to the needs and concerns of your team members. Regular one-on-one meetings can effectively understand and address individual challenges that may be absent in group settings. These meetings provide a private space for team members to discuss issues they might be facing related to the project or their work environment, which could be affecting their performance. Being responsive also means adapting your leadership style and strategies to fit the dynamics of your diverse team. For instance, while some team members might thrive under direct leadership, others might perform better with a more collaborative approach.

Managing a diverse team effectively requires a deliberate and thoughtful approach, encompassing everything from the recruitment process to day-to-day project management practices. You enhance project outcomes and contribute to a more prosperous, inclusive work environment by valuing and integrating your team members' diverse perspectives and skills. This sets the stage for successful projects and fosters a culture of respect and mutual understanding extending beyond the immediate project team to the organization.

5.6 Case Study: Turning Around a Failing Project Team

In the bustling world of project management, only some teams start on the right foot. Imagine a team embroiled in confusion and inefficiency, where missed deadlines and frayed nerves are the norms. This scenario was for a project team developing a new software tool for a large retail corporation. Initially, the team struggled significantly with poor communication, ambiguous goals, and unclear leadership, which led to low morale and subpar performance. The project needed to catch up, and stakeholders were increasingly concerned about the viability of the project outcome.

Recognizing the dire state of affairs, the company implemented decisive intervention strategies. The first step involved restructuring the team. This restructuring wasn't just about changing personnel and reevaluating and realigning the team composition to better match the project's demands. Team members with underutilized skills were given roles that better suited their strengths. New talents were brought in to fill critical gaps, particularly in areas requiring specific technical expertise.

Alongside team restructuring, there was a clear need for new leadership. While skilled in technical aspects, the existing project manager needed help with team management and communication. A new project manager with a strong background in team dynamics and a communicative leadership style was introduced. This change brought immediate relief and a fresh perspective to the team. The new leader implemented regular, structured meetings to track progress, allowing team members to voice concerns and offer solutions to ongoing issues.

The project goals were revisited and revised. Ambiguity in the original goals led to confusion and misaligned efforts. The new goals were set to be SMART—Specific, Measurable, Achievable, Relevant, and Time-bound. Clear objectives and milestones were established, giving the team a roadmap to gauge their progress and focus their efforts.

The impact of these interventions was profound. Communication within the team improved dramatically, enhancing collaboration and problem-solving. With more precise goals, team members could prioritize their tasks more effectively, leading to better time management and productivity. The restructured team leveraged their strengths, and the new leadership fostered an environment of openness and accountability.

The results spoke volumes. Not only did the team catch up to the critical milestones that were previously missed, but they also completed the project ahead of the revised schedule. The quality of work improved significantly, satisfying the rigorous requirements set by the stakeholders. The successful turnaround of the project saved the company from a potential financial disaster and restored confidence among the project team members and stakeholders.

From this experience, several vital takeaways emerge. First, aligning team members' roles with their strengths cannot be overstated. Proper alignment not only boosts individual performance but also enhances collective productivity. Secondly, leadership plays a crucial role in the health of any project. A leader who can effectively communicate, motivate, and address team issues is invaluable. Lastly, clear and achievable goals are the anchors of project success. They provide direction and a benchmark for measuring progress.

TABLE 9.1 SAMPLE RACI MATRIX

Task	Project Manager	Content Strategist	UX Designer	Graphic Designer	Front-End Developer	Back-E Develo
Project Plan	R, A	C, I	C, I	C, I	C	C
Site Map	A	C	R	I	I	I
Wireframes	A	C	R	C	I	I
Homepage Design	A	C	C	R	C	I
CMS Setup	A	I	I	I	C	R

RACI Matrix

This case study powerfully reminds us of strategic interventions' impact on a failing project team. It underscores the necessity of adaptive leadership, the strategic alignment of team roles, and the clarity of goals in rescuing and completing challenging projects. Remember these lessons as you move forward in your project management career. They are not just strategies for recovery but foundational elements that should be incorporated into every project from the outset to preempt failure and propel teams toward success.

As we conclude this chapter on leadership and team dynamics, it's important to remember that a project's strength lies not only in its team's harmony but also in the robustness of its plan. Looking ahead, the next chapter will delve into the critical aspects of project execution, where these principles of leadership, team management, and stakeholder engagement will be tested in the real-world application of project methodologies and tools, giving you the confidence to lead your team effectively.

Chapter 6

Special Topics in Project Management

In the ever-evolving landscape of project management, focusing on sustainability has become a preference and necessity. As project managers, your role in delivering successful outcomes and ensuring that these outcomes are achieved sustainably is increasingly crucial. This chapter takes you into the realm of sustainable project management, where the goal is to meet present project requirements without compromising the ability of future generations to meet their own.

6.1 Sustainable Project Management: Principles and Practices

Introduction to Sustainability in Project Management

Sustainable project management is an ethical choice and a strategic imperative in today's global environment. By integrating environmental, socio-economic, and organizational aspects into project planning and execution, this approach ensures that project outcomes contribute positively to the environmental and social systems and achieve economic gains. In a world where resources are becoming scarcer, and stakeholder expectations are growing, sustainable project management empowers you, the project manager, to create a better future and add significant value to your projects.

Sustainability in project management is not a reactive measure to comply with environmental regulations or corporate social responsibility initiatives. It's about embedding sustainable practices throughout the project lifecycle and taking a proactive stance. This proactive incorporation enhances the project's value, mitigates risks, strengthens stakeholder relationships, and builds a resilient reputation in the marketplace, making you, the project manager, a forward-thinking leader.

Integrating Sustainability: From Initiation to Closure

The integration of sustainability begins at the very inception of the project. During the initiation phase, sustainability goals should be clearly defined and aligned with the project's objectives. This alignment ensures that a clear sustainability vision guides the project team.

Planning the project provides a broader scope to embed sustainable practices. Use tools like Environmental Impact Assessments (EIA) to evaluate the potential environmental impacts associated with the project and develop strategies to mitigate these impacts. Sustainability audits, which are comprehensive reviews of a project's sustainability performance conducted at various project stages, can help monitor the implementation of sustainable practices and ensure compliance with set sustainability goals. These audits typically involve a systematic examination of the project's environmental, social, and economic performance and may include site visits, interviews with stakeholders, and a review of project documentation.

The project's execution and closure phases should continue the commitment to sustainability. During execution, sustainable resource management should be a priority. This involves efficiently using resources and minimizing waste through recycling and renewable resources. As the project nears completion, a sustainability report can be a valuable tool to evaluate how well the sustainability objectives were achieved and document the lessons learned for future projects.

Sustainable Resource Use

Effective resource management is pivotal in sustainable project management. Techniques such as using renewable resources and minimizing waste reduce the project's environmental footprint and lead to cost savings. For instance, using solar panels on site can reduce dependence on non-renewable energy sources and decrease long-term energy costs.

Additionally, employing green technology and practices can enhance the project's sustainability. Innovations such as green building materials, energy-efficient appliances, and water-saving plumbing systems contribute to sustainability and appeal to environmentally conscious stakeholders and customers.

Case Studies on Sustainable Projects

To illustrate the application of these principles, consider the case of a large-scale construction project that incorporated comprehensive sustainability practices throughout its development. The project team used locally sourced materials to reduce transportation emissions and implemented a waste management system that recycled 75% of on-site waste. The project achieved its sustainability targets and received a green certification, enhancing its market value and stakeholder satisfaction. Another example is a software development project aimed to reduce its carbon footprint. The project team adopted a virtual collaboration model, significantly reducing travel and office space usage. Advanced cloud technologies were utilized to ensure efficiency and reduce energy consumption. The project met its environmental goals and demonstrated how IT projects can contribute to corporate sustainability objectives.

Another example is a software development project aimed to reduce its carbon footprint. The project team adopted a virtual collaboration model, a work arrangement that allows team members to work together from different locations, significantly reducing travel and office space usage. Advanced cloud technologies, which enable the storage and access of data and applications over the internet, were utilized to ensure efficiency and reduce energy consumption. The project met its environmental goals and demonstrated how IT projects can contribute to corporate sustainability objectives.

These case studies highlight how integrating sustainability into project management can lead to successful, environmentally responsible outcomes while aligning with business objectives. They serve as a testament that sustainability, when embedded into project management practices, can transform the traditional project delivery approach into a forward-thinking, environmentally conscious one.

As we dig deeper into the specific topics in project management, the focus on sustainability provides a framework for achieving project success in the present and for paving the way for future environmentally sound and socially responsible initiatives. This approach ensures that as project managers, you are not just leaders in your field but also accountable stewards of the environment and committed advocates for sustainable development.

6.2 Ethical Considerations in Project Management

Navigating the complex world of project management often brings various ethical challenges that require a deep understanding and commitment to fundamental moral principles such as integrity, fairness, transparency, and responsibility. These core ethics form the bedrock upon which sustainable management practices are built, ensuring that project outcomes achieve their goals, maintain moral integrity, and promote stakeholder trust. Integrity in project management means adhering to an ethical code that fosters trust and respects all participants involved in the project. Fairness involves equitable treatment of all stakeholders, ensuring no one is unduly favored or disadvantaged. Transparency is about open communication, providing stakeholders with clear and accessible information throughout the project lifecycle, thus fostering a climate of trust and accountability. Lastly, responsibility entails acknowledging the implications of one's actions on the project and its stakeholders and being accountable for the outcomes.

Ethical dilemmas in this field are as varied as the projects themselves. For instance, a project manager might face a conflict of interest when a close friend's company bids for a project, potentially influencing their decision-making. Confidentiality issues, such as when a project involves patient data in healthcare initiatives, also frequently arise. The challenge here is to maintain the confidentiality of the information while ensuring that all necessary parties have access to the data they need to fulfill their roles. Resource misallocation is another ethical challenge where resources may not be used efficiently or are diverted to areas of the project that serve individual interests rather than the project's objectives. Other common ethical dilemmas include balancing the needs of different stakeholders, ensuring fair competition in procurement processes, and managing conflicts within the project team.

Addressing these ethical dilemmas is not just a task but a crucial responsibility for project managers. A clear framework for ethical decision-making is the key to this. It guides project managers in evaluating their choices and actions based on organizational values and professional standards. This practical approach identifies the ethical issue and considers how it impacts all stakeholders. The next step is to evaluate the options available, weighing them against ethical principles and the potential outcomes for all involved. This process should include consultation with all affected parties, ensuring that different perspectives are considered and respected. Finally, the decision should be made transparently, with a rationale that can be openly shared and justified. This process not only helps resolve ethical dilemmas but also strengthens the ethical standing of the project management practice, making each project manager a valued guardian of ethical standards.

Promoting an ethical culture within project teams and organizations is not just a task but a transformative initiative. By implementing training programs focusing on ethical decision-making and the specific ethical challenges common in project management, you can inspire a solid moral foundation among team members and leaders. Clear policies that outline expected behaviors and the procedures for handling ethical violations are also vital. These initiatives inspire a culture of integrity and respect, motivating everyone to uphold these values in their work.

Leadership by example is the most powerful tool for promoting an ethical culture. When leaders consistently demonstrate ethical behavior and decision-making, they set a standard for the rest of the team to follow. Their commitment to ethics becomes a part of the team's ethos, permeating project practices and stakeholder interactions. This leadership approach fosters an ethical culture and builds a reputation of integrity and trustworthiness in the broader community and marketplace. Leaders should also actively encourage and reward ethical behavior, creating a positive reinforcement loop that strengthens the moral culture.

Incorporating these ethical principles and frameworks into your project management practices ensures that you are not just achieving project goals but doing so in a way that upholds the highest standards of integrity and respect for all stakeholders. This ethical commitment is essential in today's globalized and interconnected world, where the impacts of project management extend far beyond immediate project outcomes, influencing broader societal and environmental systems. By effectively managing ethical dilemmas, you can avoid pitfalls, enhance stakeholder trust, foster a positive team culture, and contribute to your project's overall success and sustainability.

6.3 The Impact of Globalization on Project Management

Globalization has reshaped the landscape of project management, introducing complex challenges and opportunities that require a deft touch and a broadened perspective. While it offers access to a global talent pool and new markets, it also brings risks. For instance, managing projects that span different countries can be complicated by varying legal and regulatory frameworks. Geopolitical risks such as trade disputes or regulatory changes can impact project timelines and costs. Understanding and mitigating these risks is crucial for successful global project management.

One of the primary challenges presented by globalization is the management of diverse teams. These teams often bring diverse cultural backgrounds, languages, and professional practices. While this diversity can be a significant asset in fostering innovation and providing local insights, it also introduces communication and team dynamics complexities. Misunderstandings can arise from simple language barriers or more profound cultural differences in work habits, decision-making processes, and conflict-resolution strategies. Additionally, coordinating across time zones can complicate team interactions and project scheduling, requiring project managers to be flexible and innovative.

Global supply chains also pose unique challenges. Managing suppliers and stakeholders across different countries can introduce variability and unpredictability in quality, delivery times, and compliance with international standards. Additionally, geopolitical risks such as trade disputes or regulatory changes can impact project timelines and costs. These factors necessitate a robust risk management strategy that includes thorough due diligence, continuous monitoring, and the flexibility to adapt strategies as circumstances evolve.

Despite the challenges, globalization also offers opportunities for personal growth and learning in project management. Access to a global talent pool allows project managers to source the best expertise regardless of geographical boundaries. This access can lead to higher-quality outcomes and innovations that might not be possible in a more homogenized team. Also, international projects often provide valuable learning experiences and exposure to new markets, opening up further business growth and development opportunities. This optimistic outlook on globalization can inspire project managers to embrace the challenges and seize these opportunities for personal and professional growth.

Managing Cross-Cultural Teams

Effectively managing cross-cultural teams is a challenge and a crucial skill in global project management. It begins with an understanding and appreciation of cultural differences. Cultural sensitivity training is invaluable and essential here, providing team members with the knowledge and skills to interact respectfully and effectively with colleagues from different cultural backgrounds. In addition, establishing a culture of open communication and encouraging team members to share their cultural perspectives and experiences can foster a more inclusive and productive team environment.

Communication strategies also need to be carefully considered. Establishing clear, simple communication protocols can help minimize misunderstandings. This might include specifying the language to be used in project communications, which is often English, but also making provisions for translation or clarification when necessary. Regular virtual meetings and updates can help keep the team aligned. Still, these should be scheduled considering different time zones to ensure no one is consistently disadvantaged by attending meetings outside of regular working hours.

When it comes to conflict resolution in cross-cultural teams, a nuanced and informed approach is critical. Recognizing that diverse cultures may have unique ways of expressing disagreement and resolving conflicts is crucial. Some cultures may lean towards confrontation and clear articulation of issues, while others may prefer an indirect approach, prioritizing harmony and face-saving. Understanding and respecting these differences can pave the way for effective, culturally sensitive conflict resolution strategies, fostering healthier team dynamics.

Global Standards and Practices

Adhering to global standards and practices is another critical aspect of managing international projects. Standards such as ISO 9001 for quality management or ISO 31000 for risk management provide frameworks that can help ensure consistency and quality across projects, regardless of location. Familiarity with these standards allows project managers to implement internationally recognized best practices, enhancing the credibility and reliability of the project outcomes. Moreover, adherence to these standards can also facilitate smoother collaboration with international partners and stakeholders, as they provide a common language and understanding of project management processes.

Alongside adherence to global standards and practices, a deep understanding of local regulations and practices is vital in managing international projects. This may involve compliance with local labor laws, environmental regulations, or industry-specific standards. Often, this requires close collaboration with regional partners or consultants who can offer invaluable insights and guidance on navigating the local regulatory landscape, ensuring compliance and successful project execution.

Case Examples of Global Projects

Consider the example of a multinational corporation implementing a new IT system across its offices in Asia, Europe, and the Americas. The project team comprised members from each region, bringing diverse perspectives but also significant challenges in terms of communication and work style preferences. The project manager implemented a series of integration workshops focused on building a shared understanding of the project goals and fostering interpersonal relationships among team members. Regular rotation of meeting times ensured that no single team bore the brunt of inconvenient meeting hours, promoting fairness and consideration.

Another example involves a global health initiative to enhance healthcare delivery in several African underserved regions. The project faced numerous challenges, including logistical issues, language barriers, and varying local healthcare practices. The project management team worked closely with local healthcare providers and community leaders to tailor their approaches to the local context. They also established a centralized database accessible to all stakeholders, ensuring transparency and consistent access to project data and resources.

These examples vividly illustrate the complexities and the immense rewards of managing global projects. They highlight the need for cultural sensitivity, strategic planning, and adaptation of management practices that respect local contexts while maintaining international standards. As globalization continues to influence project management, successfully navigating these complexities can lead to a profound sense of accomplishment, offering opportunities for growth, learning, and significant contributions to global development. This should inspire and motivate project managers to take on international projects confidently and enthusiastically.

6.4 Managing Healthcare Projects: Unique Challenges and Solutions

Healthcare projects stand distinctively in project management, not merely because of their direct impact on human lives but also due to the intricate web of regulatory compliance, diverse stakeholder interests, and the critical nature of their outcomes. In managing healthcare projects, the stakes are exceptionally high, and the room for error is minimal, which naturally brings a unique set of challenges and complexities. This should underline the sense of importance and gravity in the work of healthcare project managers, reminding them of the significant impact they have on public health.

One of the most defining characteristics of healthcare projects is the stringent regulatory framework within which they operate. Whether dealing with the FDA approvals for new drugs, adhering to HIPAA regulations for patient data, or complying with local healthcare laws, each regulatory layer adds complexity to project management. Navigating these regulations involves deeply understanding legal requirements, meticulous documentation, and continuous liaison with regulatory bodies. Effective management of these regulatory aspects is crucial not just for the legal execution of the project but also for maintaining the trust and safety of the patients and stakeholders involved.

Healthcare projects typically involve a complex array of stakeholders, including but not limited to medical professionals, patients, healthcare providers, insurance companies, and regulatory authorities. Each stakeholder group has its own set of expectations, requirements, and influences, which can sometimes be conflicting. Managing these relationships and aligning their interests with the project goals requires a high degree of diplomacy, communication skills, and strategic stakeholder engagement plans. For instance, while doctors might be primarily concerned with the efficacy and safety of a new treatment, insurance companies might focus on the cost-effectiveness and compliance aspects. Balancing these diverse expectations while keeping the project on track demands robust management skills and an in-depth understanding of the healthcare ecosystem.

The critical nature of project outcomes in healthcare further amplifies the complexity of these projects. Unlike many other fields, healthcare project results can directly affect patient well-being and quality of life. This elevates the importance of precision, quality control, and ethical considerations in every project phase. Project managers in this domain must ensure that all project deliverables meet the highest standards of quality and ethical practice, which necessitates rigorous testing, validation, and continuous improvement processes. This emphasis on precision and ethics should instill a strong sense of responsibility in healthcare project managers.

Regulatory Compliance and Risk Management

Navigating the maze of regulatory compliance in healthcare projects is a daunting task that requires meticulous planning and strategic foresight. For instance, gaining FDA approval for a new pharmaceutical product involves a series of phased trials, each of which must be meticulously planned, executed, and documented. This process tests the product's efficacy, safety, and potential market impact, requiring a broad spectrum of research and risk management strategies.

Risk management in healthcare projects goes beyond typical project risks. It involves identifying potential risks to patient safety, data privacy (as mandated by HIPAA in the U.S.), and regulatory non-compliance. Developing a comprehensive risk management plan for a healthcare project typically involves thorough risk assessments, which include predicting and evaluating risks and implementing strategies to mitigate them. These strategies could range from internal audits and compliance checks to advanced training for project team members on the latest healthcare regulations and technologies.

Effective risk management also requires a proactive approach to potential changes in legislation or healthcare policies that could affect the project. Staying ahead of these changes and adapting project plans accordingly is crucial to avoid compliance issues and ensure that the project meets all legal and ethical standards.

Stakeholder Engagement in Healthcare

Effective stakeholder engagement in healthcare projects is pivotal and complex, given the diversity of stakeholders and the direct impact of project outcomes on public health. Engaging stakeholders effectively starts with identifying and analyzing all stakeholder groups and understanding their interests, influences, and potential impact on the project.

Engagement strategies must be tailored to the needs and characteristics of each stakeholder group. For medical professionals and regulatory bodies, engagement might involve detailed technical meetings, regular project progress, and compliance updates. For patients, it might involve more empathetic communication strategies, focus groups to gather feedback on patient care, and transparent information sharing about how the project could impact their treatment or care.

Developing trust with stakeholders in healthcare projects is crucial. It can be achieved by maintaining high transparency, regular and clear communication, and involving key stakeholders in the decision-making process wherever appropriate. For instance, involving a panel of healthcare professionals in the planning phase of a medical device development project could provide valuable insights into practical challenges and user requirements, which can significantly enhance the project's success.

Technological Integration in Healthcare Projects

The role of technology in healthcare projects has become increasingly significant, with advancements such as electronic health records (EHRs), telemedicine, and digital health solutions revolutionizing healthcare service delivery. Implementing these technologies in healthcare projects requires careful planning, stakeholder engagement, and compliance with health data regulations.

EHR systems, for example, need to be functional, user-friendly, and compliant with data protection regulations. Integrating EHR systems involves carefully migrating existing patient records, training medical staff, and continuous monitoring and troubleshooting to ensure the system operates smoothly and securely.

Conversely, telemedicine projects involve unique challenges, such as ensuring the reliability of digital communication tools, maintaining patient privacy, and integrating with different healthcare systems. These projects often require innovative solutions to ensure that remote healthcare services are as effective and reliable as traditional in-person consultations.

Integrating technology in healthcare projects improves efficiency and patient care and requires a forward-thinking approach to project management that embraces innovation, compliance, and user-centric design.

6.5 The Construction Project Lifecycle: A Detailed Guide

Navigating the construction project lifecycle requires a strategic and well-orchestrated approach from the initial conception to completion. Each phase has unique activities and challenges that require meticulous attention and management expertise. Let's break down these phases to understand better what to expect and how to manage each stage effectively.

The lifecycle begins with the conception phase, where the project idea is born. This phase involves preliminary planning and feasibility studies. You need to assess the project's viability in terms of environmental impact, community needs, and financial practicality. During this phase, engaging with stakeholders, such as future users, community leaders, and regulatory bodies, is crucial to gathering support and input, significantly influencing the design and planning process.

Following conception, the project moves into the detailed planning and design phase. Here, architects and engineers develop detailed blueprints and plans based on the inputs gathered during the conception phase. This stage is critical as it lays down the technical specifications and the project roadmap. Integrating eco-friendly materials and energy-efficient systems right from the start is essential for considering sustainability and efficiency in your designs. This helps reduce the environmental impact and ensures cost-effectiveness over the project's lifecycle.

The project advances to the procurement phase once the planning and designs are approved, usually by regulatory bodies and stakeholders. This phase involves selecting contractors and suppliers. It's a pivotal stage where your negotiation skills are crucial in obtaining the best materials and labor at the most reasonable costs without compromising quality. Managing timelines and ensuring all parties understand their roles and responsibilities is vital to keeping the project on track.

Construction Phase Challenges

As you transition into the construction phase, several challenges might emerge, such as weather delays, labor issues, and budget overruns, each capable of derailing your project's progress. Weather delays are unpredictable, but planning for eventualities in your project timeline can mitigate their impact. Ensure that your contracts include clauses that address such delays, providing clear guidelines for proceeding in such situations.

Labor issues, including shortages or disputes, can significantly impact your project. To manage labor effectively, maintain open lines of communication with team leaders and labor unions. Regular meetings and updates can help anticipate potential issues and address them proactively. Offering fair wages and ensuring safe working conditions are essential in maintaining a motivated and efficient workforce.

Budget overruns are another common challenge in construction projects. These can be due to various reasons, including changes in project scope, unforeseen obstacles, or increases in material costs. To manage budget overruns, keep a stringent check on expenditures, and have contingency funds in place. Regular financial audits can help identify and address discrepancies early. Engaging a skilled financial manager for your project could also provide additional oversight, ensuring that your project stays financially viable.

Quality Assurance and Control

Ensuring that your construction project meets all specified standards and regulations is non-negligible. Quality assurance and control guarantee your construction project's safety, usability, and longevity. Implementing a robust quality control system involves regular inspections and audits at different stages of the construction process. Employ experienced inspectors who can identify potential issues early in the process, allowing for timely corrections that save time and money.

Utilize technology to enhance your quality control processes. Advanced software can help track the quality of materials and workmanship throughout the construction process, ensuring compliance with the project specifications. Regular training sessions for your team on quality standards and expectations also play a crucial role in maintaining high standards throughout construction.

Also, adhering to regulatory standards is about compliance and ensuring the safety and well-being of the end-users and the community. Regular liaisons with regulatory bodies and adherence to safety and building codes are essential. This helps avoid legal complications and builds trust and credibility with stakeholders and clients.

As you manage your way through these complex layers of the construction project lifecycle, remember that each phase offers unique opportunities for innovation and improvement. Your role as a project manager is to steer these opportunities in a direction that ensures the project not only meets its intended goals but does so in an efficient, sustainable, and exemplary.

6.6 Financial Project Management: Budgeting and Cost Control

Navigating through the financial aspects of project management is akin to steering a ship through a sea of fluctuating markets and unforeseen costs. It demands a keen eye on budgets, forecasts, and financial reports, ensuring that every dollar spent is an investment towards the project's success. Understanding the fundamental principles of financial management within the context of projects is crucial. These principles encompass budgeting, forecasting, and financial reporting—each serving as a cornerstone to maintaining the economic health of your projects.

Budgeting is the blueprint of your financial strategy. It involves detailing the estimated costs associated with every aspect of the project—from initial resources and labor to ongoing operational expenses. Creating an accurate project budget requires a thorough understanding of the project's scope and the resources needed. It's about forecasting costs realistically, considering potential financial risks, and preparing for them. This process isn't just about numbers; it's about setting expectations, defining limitations, and establishing a financial roadmap that aligns with your project goals.

Once the budget is set, managing it effectively becomes your next challenge. This is where techniques for cost estimation and allocation play a pivotal role. Cost estimation involves analyzing different aspects of the project to predict financial needs accurately. Tools and methods like analogous estimating, parametric modeling, and bottom-up estimating can be employed depending on the project's complexity and the accuracy of the data available. Allocation is about assigning the estimated costs to different parts of the project, ensuring that funds are available where and when needed most, thus avoiding financial bottlenecks that could delay project progress.

Cost Control Techniques

Maintaining control over your project's finances requires more than just careful planning—it requires continuous monitoring and adjustment. Cost control techniques such as Earned Value Management (EVM), variance analysis, and cost performance indexes (CPI) are instrumental in this ongoing process. EVM, for instance, offers a comprehensive method that integrates the project scope, schedule, and cost elements to help project managers assess project performance and its alignment with the financial plan. Through EVM, you can identify cost variances between the planned and actual expenditures and take corrective actions timely.

Variance analysis further supports cost control by highlighting deviations from the budget. This technique allows you to dig deeper into each variance, understanding its cause and impact on the project. Whether it's due to unexpected external factors or errors in initial estimates, identifying these variances early helps you mitigate financial risks effectively. Meanwhile, the Cost Performance Index quantitatively measures cost efficiency, comparing the budgeted cost of work performed to the actual price. This metric is invaluable for assessing current performance and forecasting future financial needs, ensuring the project stays on budget.

Financial Reporting and Stakeholder Communication

Transparent and effective communication of the project's financial status is crucial in maintaining stakeholder trust and engagement. Financial reporting is not just a statutory requirement or a routine task; it's an essential communication tool that informs stakeholders about how their investments are being managed. Regular financial reports should provide clear, concise, and accurate information about the project's financial health, offering insights into how funds are being spent, the efficiency of the financial management practices in place, and the financial forecast for the project.

These reports serve as a basis for financial decision-making, helping stakeholders understand the project's financial dynamics and facilitating informed discussions about future financial planning. Additionally, the financial reporting process keeps the project team accountable, ensuring that every economic decision aligns with the project's objectives and compliance requirements.

Effective financial project management is about much more than keeping the books balanced. It's about strategic planning, rigorous monitoring, and clear communication, all essential to steering your project to financial and operational success. As you apply these principles and techniques, remember that every financial decision you make directly impacts the project's outcome. By mastering financial project management, you safeguard your project's assets and contribute to its overall success, ensuring it delivers value to all stakeholders.

As this chapter concludes, remember that financial management is a critical skill in the toolkit of successful project managers. The principles and practices discussed here are fundamental to navigating the economic complexities of modern projects and ensuring their success. Looking ahead, the next chapter will explore another vital aspect of project management, preparing you to expand your expertise further and enhance your ability to lead projects effectively.

Chapter 7

Preparing for Certification and Career Advancement

Envision the profound sense of accomplishment and the vast array of new opportunities that come with holding a project management certification. These certifications, such as PMP, CAPM, and Agile credentials, are not mere pieces of paper. They are powerful symbols of your unwavering dedication and unparalleled expertise, battle shields, and swords that equip you to stand out and triumph in the fiercely competitive field of project management.

7.1 Overview of Project Management Certifications: PMP, CAPM, and Beyond

Certification Options

Navigating the myriad of project management certifications can feel like exploring a dense forest, each path promising to lead to a new peak. Among the most recognized certifications is the Project Management Professional (PMP)® credential, offered by the Project Management Institute (PMI)®. It is a gold standard in the industry, symbolizing expertise and commitment to the profession. The Certified Associate in Project Management (CAPM)® provides a valuable stepping stone for those newer to the field, offering a foundational understanding of key management processes and terminology.

Beyond these, if your projects are taking on an Agile flavor, considering Agile-specific certifications such as Certified ScrumMaster® (CSM) or PMI-Agile

Certified Practitioner (PMI-ACP)® could be beneficial. These certifications enrich your knowledge pool and align you with current global methodologies that are reshaping industries.

Certification Requirements

Each certification has its own set of prerequisites and paths. For instance, to sit for the PMP® exam, you need either a secondary degree with 7,500 hours leading projects and 35 hours of project management education or a four-year degree with 4,500 hours leading projects and the same amount of education. The CAPM®, however, requires a secondary diploma and either 1,500 hours of project experience or 23 hours of project management education before you sit for the exam.

The application process generally involves demonstrating your eligibility through detailed documentation of your education and experience, followed by an exam that tests your knowledge and application of project management principles. These exams are meticulously crafted to cover all aspects of project management, ensuring a comprehensive assessment of your capabilities.

Benefits of Certification

Acquiring a project management certification can be a transformative catalyst for your career. It's not just about validating your skills and knowledge; it's about unlocking doors to new job opportunities and significantly boosting your earning potential. Research has shown that individuals with certifications like PMP® can earn over 20% more than their non-certified counterparts. So, envision the myriad of possibilities with a project management certification.

Choosing the Right Certification

Choosing the most suitable certification is a pivotal decision that can profoundly shape your career trajectory and potential earnings. It should be a well-informed choice based on your career goals, experience level, and the specific demands of your industry. If you're starting, the CAPM® might be the perfect choice, providing you with the credentials to progress to more complex projects. A specialized Agile certification could be more advantageous if deeply immersed in Agile project environments.

For those aiming to solidify their expertise and broaden their career opportunities in various industries, the PMP® is an excellent choice. It enhances your professional growth and equips you with skills applicable across numerous sectors, from IT to construction.

Interactive Element: Reflection Section

Think About Your Certification Goals

- **What are your long-term career objectives?**
- **How can certification help you achieve these goals?**
- **Which certification aligns best with your current skills and future ambitions?**

Reflecting on these questions can help clarify your path toward certification and ensure that your choice today aligns with the career you envision for tomorrow.

Embarking on the certification path is not just about adding credentials to your resume; it's about making a strategic move toward a more prosperous and fulfilling career in project management. By understanding the different certifications available, their requirements, and the doors they can open, you're better prepared to make an informed decision that boosts your professional life and sets you apart in the competitive landscape of project management.

In the following sections of this chapter, we will explore effective study strategies to ensure you're well-prepared for your certification exams and tips on developing a robust career path that leverages your new qualifications to their fullest potential. Stay tuned as we continue to unfold the essentials of advancing your career in project management.

7.2 Study Strategies for Passing Project Management Exams

A well-structured study plan is not just a roadmap but a key to unlocking your potential as a project manager. It guides you through the often dense and complex material covered in project management exams, leading to a sense of accomplishment when you pass. Imagine this plan as your daily itinerary on a long hike; it outlines your starting point each morning, marks the milestones you aim to reach, and ensures you have a clear direction. To construct a practical study schedule, begin by outlining the syllabus and estimating your time before the exam. Break down the material into manageable sections and allocate specific times to cover each, ensuring you have time for thorough review and practice exams. Daily goals can transform overwhelming content into achievable tasks, making your preparation more structured and less daunting.

Incorporating proven study techniques can significantly enhance your retention and understanding. Active recall, where you test your knowledge regularly, forces you to retrieve information from memory, strengthening your recall over time. You might use flashcards or practice questions to implement this technique. Spaced repetition leverages the psychological spacing effect, where information is reviewed at increasing intervals to embed deeply into memory. This method can be facilitated by scheduling review sessions for each topic one day, one week, and one month after you initially study it, ensuring the material is revisited just as it begins to fade from memory. Additionally, study groups can provide a multifaceted benefit. Discussing complex topics with peers can uncover new perspectives and clarify misunderstandings while teaching others is a powerful method to reinforce your understanding and identify gaps in your knowledge.

Choosing the right resources for exam preparation can substantially impact your performance. The ideal resources cover all necessary content and match your learning style. Look for books comprehensively aligned with the certification's body of knowledge, and consider online courses that offer interactive components and regular quizzes. Experienced instructor-led workshops can provide deeper insights and clarify doubts in real-time. Furthermore, simulation exams are invaluable; they not only familiarize you with the format and time constraints of the actual exam but also put your knowledge to the test under exam conditions, which can help identify areas that need more attention.

Handling exam anxiety is a crucial aspect of your preparation. It's natural to feel nervous, but unchecked anxiety can undermine months of hard work. Establish a regular study routine to build confidence through familiarity with the material. Techniques such as mindfulness and deep breathing exercises can effectively manage stress. Instead of cramming on the day before the exam, focus on relaxing and organizing your materials for the exam day. Ensure you have a good night's sleep; rest is crucial for cognitive function and memory recall. During the exam, manage your time wisely, and if anxiety strikes, pause for a moment, take a few deep breaths, and refocus on the questions. Remember, preparation is critical, and each step in your study plan is a step toward success.

By implementing these strategies, you equip yourself to pass the exam confidently. Remember, the goal is to gain a certification and deeply understand and effectively apply project management principles in your professional life.

7.3 Developing a Career Path in Project Management

Pursuing a fulfilling career in project management is about laying the foundations and building a future that aligns with your skills, strengths, and interests. This process is akin to laying the foundations for a building; ensuring that your career is built on a solid understanding of what you excel at and enjoy is crucial. Begin by listing your current skills and experiences, then rate how proficient you are at each and how much you enjoy each activity. This exercise highlights your strengths and pinpoints the areas where your passion and skills intersect, leading to more satisfying career development.

Once you clearly understand your skills and interests, the next step is to visualize your career trajectory through career mapping. Think of this as drawing a treasure map where X marks your ultimate career goal. Start by identifying where you are now and where you want to be. Do you see yourself leading large-scale international projects, or are you more interested in the agile startup environment? Mapping out your career involves plotting intermediate positions as stepping stones toward your ultimate goal. Each position should help you acquire skills and experiences that bring you closer to where you want to be. For instance, if you aim to manage large infrastructure projects, you might start as a junior project coordinator, progress to a project manager, and specialize in infrastructure projects as you gain more experience.

Developing the necessary skills is a process crucial to your chosen career path. Continuous learning is key in the ever-evolving field of project management. Focus on hard skills like risk management and quality control and soft skills like leadership and communication. Given the dynamic nature of project management, staying updated with the latest project management software and tools is also beneficial. Consider regular training sessions, certification courses, or workshops to sharpen your skills. For instance, understanding the latest project management software, like JIRA or Asana, could be invaluable if you're inclined toward technology projects.

In addition to skill development, staying abreast of industry trends is not just a necessity but a powerful tool that empowers you. The landscape of project management continually evolves with new methodologies, standards, and technologies emerging. Keeping updated with these changes ensures your skills remain relevant and open new career advancement opportunities. Subscribe to industry publications, join appropriate forums, and participate in webinars and conferences. This proactive approach keeps you informed and positions you as a knowledgeable professional who is in control of your professional growth.

Empower your project management career by taking charge of your learning. Understanding your strengths and charting your career path is a proactive approach that allows you to enhance your skills and stay updated on industry trends continuously. This self-driven strategy sets the stage for a successful and fulfilling career in this dynamic field. Remember, the most successful project managers manage projects and their career development with intention and strategy.

7.4 Networking and Professional Development in Project Management

Building a solid professional network is a strategic effort that can foster community in the project management industry. As a project manager, expanding your circle and connecting with peers across the sector can lead to remarkable opportunities for growth and learning. Engaging in industry conferences is one such strategy. These gatherings provide networking opportunities and a chance to build relationships with industry leaders, potential collaborators, and like-minded professionals. By attending these events, you gain insights into the latest industry trends and innovations and get the chance to present your ideas and projects, which can enhance your professional reputation.

Joining professional associations such as the Project Management Institute (PMI) provides a structured pathway to networking. These organizations are not just about adding a credential; they are vibrant communities where ongoing professional development is encouraged through workshops, seminars, and certification courses. PMI, for instance, offers myriad resources such as webinars, local chapters, and special interest groups that allow you to dive deeper into specific areas of project management, from Agile practices to risk management. The connections in these settings often involve professionals who share a commitment to their careers and a passion for improving their craft, providing a network that can support your professional journey through advice, mentorship, and more.

The benefits of a well-maintained professional network are manifold. Networking can expose you to job opportunities that may not be publicly advertised, often called the "hidden job market." Many companies prefer to hire through referrals, and having a connection within the company can significantly boost your chances of landing an interview. Moreover, the diverse perspectives you gain from your network can inspire innovative solutions to project challenges and improve your problem-solving skills. Perhaps one of the most underrated benefits is the potential for finding mentors who can guide you based on their experiences, offering advice tailored to your career aspirations and challenges.

In today's digital age, online networking opportunities have increased, making it easier than ever to connect with professionals across the globe. Platforms like LinkedIn allow you to join project management groups where discussions on current topics can keep you informed and engaged. Participating actively in these discussions by sharing your insights and asking thoughtful questions can elevate your visibility and establish you as an astute leader in the field. Furthermore, many professional organizations host virtual meetups, webinars, and forums that provide opportunities to learn and network without traveling, ensuring you can connect with global peers from the comfort of your home or office.

Engaging effectively in these online spaces requires a proactive and strategic approach. Regular updates to your profiles with recent projects, achievements, and learnings can attract professional attention. Additionally, offering help or advice when others post queries or challenges builds goodwill and establishes your expertise in the field. The key to successful online networking is consistency; regular interactions and contributions can develop and maintain your professional presence, keeping you relevant in the rapidly evolving project management landscape.

By embracing traditional and modern networking strategies, you can develop a professional network that supports your current role and paves the way for future career opportunities. Whether through face-to-face interactions at conferences, active participation in professional associations, or engaging discussions on digital platforms, each connection you make is a step toward a more prosperous, more informed, and interconnected professional life. Thus, invest in your professional network with the same zeal and strategy as you would in a high-stakes project—it's an investment that pays dividends in countless ways, propelling your career in project management forward.

7.5 The Role of Mentorship in Project Management Career Growth

Mentorship in project management is akin to having a seasoned guide while navigating a complex trail; it can make the journey smoother and more insightful. Finding the right mentor, however, requires thoughtful consideration and a strategic approach. Start by identifying what you seek in a mentor—someone whose career path aligns with your aspirations or someone known for their expertise in a particular aspect of project management that interests you. Professional settings, industry conferences, and LinkedIn are excellent places to look for potential mentors. Once you identify a potential mentor, the approach should be respectful and transparent. Please introduce yourself, explain why you value their expertise, and express your desire for guidance. Remember, showing genuine interest and respect for their time and knowledge is vital.

The benefits of having a mentor in project management are not just substantial, but they provide guidance and support. A mentor provides more than just knowledge; they offer nuanced insights that only come with experience. This guidance can help you navigate complex project challenges, avoid common pitfalls, and make more informed decisions. Remember, mentors often introduce mentees to their professional network, significantly expanding your connections and exposing you to new opportunities. This networking aspect can be invaluable as you look to advance your career. Mentors often provide emotional support, encouraging you during tough times and celebrating your successes. Their belief in your capabilities can boost your confidence, driving you to take on and conquer more challenging roles and making you feel guided and supported in your professional journey.

Being an effective mentee is crucial to the success of the mentorship relationship. Proactivity is vital; take charge of your learning by asking questions, seeking advice, and discussing your career progression plans. Respect for the mentor's time is also critical. Be punctual for meetings, adhere to agreed schedules, and come prepared with specific topics or questions you need guidance on. Being open to feedback, including constructive criticism, is essential. Use the feedback to reflect on and improve your project management practices. This openness helps your personal growth and strengthens the mentor-mentee relationship through honest and productive exchanges.

Mentorship success stories abound in project management, each illustrating the profound impact a mentor can have on a mentee's career. Consider the case of a young project manager struggling to lead a diverse team in a highly competitive IT environment. Through mentorship, they learned effective communication strategies and leadership skills tailored to managing diverse groups. The guidance helped them steer their team more effectively and led to completing a critical project, accelerating their career progression. Another inspiring story involves a project manager who, under the mentorship of a seasoned industry expert, was able to transition from traditional project management methods to agile practices effectively. This shift enhanced their project delivery times and opened up new career opportunities in industries that value agility and innovation.

These stories highlight the tangible benefits of mentorship in providing career guidance, skill enhancement, and emotional support, ultimately fostering professional growth and success. As you consider integrating mentorship into your career development strategy, consider it an investment in your future that provides immediate assistance and long-term career benefits. Engaging with a mentor can be one of the most impactful steps you take on your career path, providing you with the insights and support needed to navigate the complexities of project management and achieve your professional goals.

7.6 Case Study: A Project Manager's Career Advancement Journey

In the bustling world of project management, the career trajectory of a successful project manager often serves as a blueprint for aspirants aiming to carve their niches in this dynamic field. Let's consider the journey of Sam, a project manager whose career blossomed from humble beginnings into a role model of success and resilience.

Sam's career began at a small software development firm, handling minor projects with limited scope. The initial challenges were daunting—tight budgets, demanding clients, and a team skeptical of a young manager's directives. What set Sam apart was his willingness to dive deep into the nuances of each project, ensuring that no detail was overlooked. His breakthrough came when he led a project implementing a new software deployment, significantly enhancing the client's operational efficiency. This success was pivotal; it boosted his confidence and cemented his reputation as a capable leader within his firm.

Strategic career moves significantly shaped Sam's professional journey. After his early success, he pursued a PMP certification, realizing that formal validation of his skills would open up new avenues. The certification process was rigorous, but it equipped him with a comprehensive understanding of advanced project management principles, which he could apply to larger, more complex projects. Another strategic decision was his shift to a multinational corporation, where he managed cross-functional teams and multi-tiered projects. His desire for broader exposure and more challenging environments fueled this move. Here, Sam excelled by leveraging his deep knowledge and adaptive strategies to steer complex projects involving advanced technologies and diverse teams.

Sam learned valuable lessons throughout his career that shaped his management style and professional outlook. One key lesson was the importance of clear communication. Early in his career, a project faltered due to misunderstandings between team members and stakeholders. This experience taught him that articulating project goals and updates clearly and regularly was crucial to aligning team efforts with client expectations. Another lesson was the significance of stakeholder engagement. Sam found that actively involving stakeholders at every project phase preempted potential issues and fostered a collaborative environment conducive to success.

Sam offers advice for those starting in project management: embrace every challenge as an opportunity to learn. Whether it's a difficult client, a failed project, or an unfamiliar task, each challenge is a stepping stone to greater understanding and skill. He also stresses the importance of continuous learning and adaptability. The project management landscape is perpetually evolving, with new methodologies and technologies constantly emerging. Staying updated and adaptable ensures relevance and competitiveness in the field.

Sam's journey is a testament to the potential of well-planned career moves and the relentless pursuit of professional growth. His story, marked by growth and success, inspires aspiring project managers. It encourages them to pursue their professional goals with determination and openness to learning, ensuring a journey that is not just about reaching milestones but about the sense of accomplishment that comes with it.

As we wrap up this chapter, it's clear that the path to becoming a successful project manager is multifaceted, involving strategic learning, career planning, and personal development. Each story, including Sam's, serves as a testament to the potential that lies in well-planned career moves and the relentless pursuit of professional growth. As you step forward, remember that each challenge is an opportunity, and every learning experience is a step toward your success in the ever-evolving world of project management. Let these stories inspire you as you continue to navigate your path in this exciting and rewarding field.

Chapter 8
The Future of Project Management

As the sun sets on traditional project management practices, a new dawn beckons—a future where the role of a project manager will not just be about managing timelines and resources but about leading strategic initiatives that drive substantial business value. Imagine the empowerment of playing a pivotal role in shaping the strategic direction of your organization, where your insights influence corporate strategy and your projects directly impact the company's bottom line. This is not just a possibility; it's the expected evolution of the project manager's role in the coming decade. Your role is crucial, and your contributions are invaluable.

8.1 The Evolving Role of the Project Manager in the Next Decade

Shift Towards Strategic Leadership

The landscape of project management is undergoing a significant transformation, moving away from traditional task-based roles and towards strategic leadership positions. The role of project managers is evolving, with a growing expectation for them to act as strategic leaders. This shift necessitates a deeper understanding of the broader business context in which projects operate. It's not just about overseeing project execution; it's about ensuring that projects align with the organization's strategic objectives. For instance, if a company aims to expand its market share, you, as a project manager, might lead the launch of new products designed to capture emerging markets, thereby directly contributing to the strategic goal of market expansion. A real-life example of this is when a project manager at a tech company led the development and launch of a new software product, which significantly increased the company's market share in a new industry.

Increasing Importance of Soft Skills

As project managers transition to strategic roles, the significance of soft skills is amplified. Skills such as emotional intelligence, negotiation, and leadership are becoming as vital as technical skills. Emotional intelligence equips you to navigate the complexities of team dynamics, fostering a collaborative work environment and enhancing team performance. Negotiation skills are crucial in securing resources, managing stakeholder expectations, and resolving conflicts. Leadership, however, is the bedrock of effective project management. It involves inspiring and motivating your team, driving change, and making decisions that impact the project's success. For example, using emotional intelligence to assess and address team concerns during a project meeting can prevent dissatisfaction and disengagement, ensuring the team remains committed and productive. Other increasingly important soft skills for project managers include communication, adaptability, and problem-solving.

Adaptation to Technological Advances

Technology advances unprecedentedly, and modern project managers must keep pace with these changes. From AI and machine learning that can predict project risks and automate routine tasks to blockchain technology that offers enhanced transparency and security, the tools at your disposal are powerful and transformative. Integrating these technologies into your project workflows can boost efficiency and give you unprecedented insights and control over your projects. For instance, you can use AI-driven analytics to predict potential schedule overruns and budget breaches before they occur, allowing for timely intervention.

Sustainability and Social Impact

In an era where sustainability and corporate responsibility are at the forefront of business practices, project managers are increasingly tasked with leading projects that deliver economic value and benefit society and the environment. This involves managing projects that adhere to sustainable practices and ethical standards and aligning project outcomes with global sustainability goals. Whether reducing waste through lean management techniques or implementing energy-efficient technologies, the focus is creating value that transcends financial gains, benefiting stakeholders and communities. For example, a project aimed at reducing water usage in manufacturing cuts costs and minimizes environmental impact, demonstrating the project's alignment with broader sustainability goals. Your projects have the power to make a difference to create a better world.

Navigating this evolving landscape offers a unique opportunity for personal growth and development. It's a journey that requires you to be proactive, continuously seeking to develop your strategic thinking, technological proficiency, leadership skills, and commitment to sustainability. As you step into the future of project management, remember that your role is expanding. You are no longer just managing tasks; you are leading initiatives that have the potential to transform your organization and make a positive impact on the world. This is your chance to grow, to evolve, and to make a difference.

8.2 Project Management Trends to Watch: From Agile to Hybrid Models

The project management landscape is continuously evolving, adapting to the shifting demands of industries and the innovations that drive them. One of the most significant shifts we are witnessing is the rise of hybrid project management methodologies. These methodologies represent a convergence of traditional practices, known for their structured approach, and Agile practices, celebrated for their flexibility and responsiveness. This blend offers a customized toolkit that allows project managers to navigate the complexities of modern projects that might not strictly adhere to one methodology over another. For instance, consider a long-term construction project where the waterfall method provides a structured approach while integrating Agile practices can address changes more dynamically during the execution phase. This hybrid approach ensures that project management is both adaptive and predictive, catering to the detailed-oriented scope of traditional methods with the flexibility to incorporate changes swiftly and efficiently as they arise.

Expanding the Agile methodology beyond its IT genesis into sectors like manufacturing and healthcare is also reshaping how projects are managed across different industries. Agile can streamline manufacturing production processes by improving feedback loops and accelerating response times to market changes. However, the challenge lies in integrating Agile in a highly structured environment where changes can be costly and slow to implement. Despite these challenges, the benefits—such as increased adaptability and enhanced collaboration—often outweigh the hurdles. For instance, Agile implementation in a manufacturing project led to a significant reduction in production time and cost despite the initial challenges of adapting to a new methodology. Similarly, Agile methodologies improve patient care services by developing processes that adapt quickly to the changing healthcare landscape. Implementing Agile in these sectors involves a shift in project management practices and a cultural shift within organizations to embrace flexibility over structure.

The growing use of data analytics in project management is another trend transforming the field. Data analytics enables project managers to make more informed decisions by predicting project outcomes, identifying risks, and optimizing resources. With the ability to analyze vast amounts of data in real time, managers can foresee potential delays or budget overruns and mitigate them before they escalate. For example, data analytics can reveal that the procurement process is the bottleneck causing project delays, allowing managers to prioritize improvements in that area. The predictive power of data analytics enhances decision-making and improves projects' overall efficiency and success rate. A practical application of this is when a project manager uses data analytics to identify potential risks in the project's timeline, allowing them to adjust the schedule and prevent a delay.

A focus on customer-centric project management is becoming increasingly prevalent, particularly in industries where customer satisfaction is directly linked to business success. This approach prioritizes the needs and expectations of the customer throughout the project lifecycle, from planning and development to delivery and feedback. In product development, for example, a customer-centric approach means continuously engaging with the customer to ensure the product meets their needs and adjusting the project scope based on their feedback. This can lead to more successful products and higher customer satisfaction, as user needs and preferences shape the final product. Adopting a customer-centric approach requires a deep understanding of one's customer base and the flexibility to adapt project goals and processes according to evolving customer insights.

These trends underscore the need for project managers to be versatile, continually learning, and adept at applying various methodologies and tools to meet their projects' unique demands. As industries continue to evolve and new technologies emerge, the ability to integrate traditional and Agile methods, leverage data analytics, and maintain a customer-centric focus will be vital to managing successful projects in the future.

8.3 Continuous Improvement and Learning in Project Management

In the fast-paced realm of project management, the only constant is change. Staying updated with the latest industry standards and practices is beneficial and essential for maintaining competitive edge and operational effectiveness. This necessity for lifelong learning isn't just about individual growth—it's about ensuring your projects and team remain at the forefront of efficiency and innovation. Picture yourself not just keeping up with trends but setting them up and becoming a beacon of knowledge and adaptability in your organization.

Integrating learning directly into the work environment can transform the traditional approach to professional development. Imagine real-time training sessions that address immediate project challenges or digital learning platforms that provide on-demand resources tailored to specific project needs. This approach ensures learning is relevant and immediately applicable, enhancing retention and engagement. Collaborative learning environments further enrich this experience, allowing team members to share insights and solutions, turning every project into a learning opportunity. For instance, after-action reviews at the end of project phases can be powerful learning sessions where team members discuss what went well and what could be improved, ensuring lessons are learned and applied in future projects.

Adaptive learning technologies are revolutionizing how project managers upskill. These technologies use algorithms to tailor learning content to the user's pace and performance, addressing individual knowledge gaps and learning preferences. For you, this could mean engaging with an interactive module on risk management that adapts its complexity based on your responses, ensuring you are neither bored with simplicity nor overwhelmed by complexity. This personalized approach not only makes learning more effective but also more engaging, encouraging continuous professional development.

Fostering an organization's culture of improvement and learning is a beneficial practice that can significantly enhance project outcomes. Encouraging innovation and flexibility, this culture supports risk-taking and iterative improvement, essential components of modern project management. You can lead by example, championing learning initiatives and celebrating innovative solutions that enhance project efficiency—promoting an environment where feedback is not just valued but actively sought after. Failures are not just seen as learning opportunities but as crucial stepping stones that can transform the organizational mindset, embedding continuous improvement into the corporate DNA. For example, regular brainstorming sessions that welcome ideas from all team members, regardless of their role or seniority, can spur innovation, driving the development of new strategies and solutions that keep your projects—and your team—ahead of the curve.

In this ever-evolving landscape, your role as a project manager transcends managing tasks and timelines. You are not just a manager but a facilitator of learning and an agent of change, driving project success and your team's continuous growth and adaptation. Embrace this role with enthusiasm and commitment, and watch as it transforms not only your projects but also your career, positioning you as a leader in project management. Your ability to inspire and motivate your team to embrace change and learning is key to your success as a project manager.

8.4 The Integration of Cross-Cultural Competences in Global Projects

Managing global projects across different cultures, time zones, and regulatory environments is becoming increasingly critical in today's interconnected world. This global landscape presents a unique set of challenges and opportunities for project managers. As projects stretch across borders, they often bring together a diverse mix of individuals who bring their cultural perspectives, working styles, and expectations to the table. Effectively Navigating this diversity is necessary and a significant advantage in the global marketplace.

Developing cross-cultural competence involves more than just acknowledging cultural differences—it requires active engagement and a deep understanding of how these differences can influence project outcomes. For instance, communication styles vary widely across cultures: some may prefer direct communication, while others value a more indirect approach. Understanding these nuances and adapting your communication style can prevent misunderstandings and foster a more collaborative atmosphere. Similarly, negotiation practices differ across cultures—what is considered a persuasive argument in one culture may be seen as aggressive or rude in another. Developing a keen awareness of these differences can tailor your approach to the cultural context, ensuring more effective interactions and negotiations.

Another critical aspect is adapting project management practices to align with local regulations and cultural expectations. This might include modifying standard procedures to comply with local labor laws or adjusting project timelines to accommodate local holidays and work schedules. Such adaptations ensure compliance and demonstrate respect for the local team's culture and practices, which can enhance team cohesion and project commitment.

Building a successful global project team also involves overcoming language barriers and differing work ethics. Consider using translation services or multilingual team members to bridge language gaps and ensure clear and accurate communication. Establishing clear, written communications can also help minimize misunderstandings. Regarding work ethics, setting clear expectations and building mutual respect are essential. Recognizing and accommodating different work approaches is important while fostering a shared commitment to the project's goals. Regular virtual meetings and team-building exercises can help strengthen relationships and align diverse team members toward common objectives.

Numerous success stories well illustrate the rewards of effectively managing cross-cultural teams. One notable example involved a multinational technology firm that launched a complex software development project involving teams in India, Germany, and the United States. Despite initial challenges related to time zone differences and communication barriers, the project manager implemented regular cross-cultural training sessions and established a 24-hour project management cycle to ensure continuous progress and communication across all teams. By fostering an inclusive culture and adapting management practices to accommodate the team's diverse needs, the project was completed ahead of schedule and under budget, resulting in significant cost savings and enhanced client satisfaction.

As you navigate projects that span different geographical and cultural boundaries, remember that understanding, respecting, and integrating diverse cultural perspectives is not just a challenge, but also a source of personal and professional growth. It enhances project efficiency and outcomes and enriches your professional experience, preparing you for the increasingly global nature of project management. The successes you achieve in this realm will underscore the empowerment that comes with understanding cross-cultural competencies in international project management.

8.5 Innovations in Project Documentation and Reporting

In the dynamic landscape of project management, digital transformation is revolutionizing how documentation is created, stored, and shared. Embrace the shift to cloud-based platforms, where documents are stored securely and accessible in real-time to everyone on the team, regardless of their physical location. This accessibility is crucial in today's globalized work environment, where team members work across different time zones and locations. Imagine the efficiency gained when updates to critical project documents are instantly available to all stakeholders, eliminating the delays that once hampered project progress. Remember, these platforms often come with advanced features such as version control, which ensures that everyone is working from the most current document, reducing confusion and errors from using outdated information. These innovations are about technology and empowering you to be more productive and effective in your role.

The rise of automated reporting tools represents another significant advancement in project management. These tools can automatically gather data from various sources within the project management software, providing real-time insights into project status, resource allocation, and performance metrics. For example, they can collect data on tasks completed, hours worked, or budget spent. Automating these processes saves time and increases accuracy, as the potential for human error in data collection and reporting is significantly reduced. For instance, consider a tool that can instantly generate a report showing the current expenditure of the project against the budget. This capability allows for swift decision-making and adjustments to keep the project within financial boundaries. The development and implementation of these tools are becoming increasingly sophisticated, with features that allow for customized reports tailored to the specific needs of stakeholders, ensuring that all team members have the exact information they need when needed.

Enhancing transparency with stakeholders is another area where innovations in documentation and reporting are making a significant impact. The ability to provide stakeholders with up-to-date, accessible, and accurate information builds trust and fosters a collaborative relationship. This transparency is crucial, especially in projects where stakeholders make significant financial or operational commitments. For instance, if a new regulation requires projects to report their environmental impact, transparency about your project's compliance efforts can help build stakeholder trust. Project managers can use digital platforms and automated tools to continually inform stakeholders about the project's progress, challenges, and successes. This ongoing engagement keeps stakeholders informed and enables them to provide timely feedback that can influence project direction, ensuring that the project aligns closely with their expectations and needs.

Looking toward the future, the role of predictive analytics and artificial intelligence (AI) in project reporting is poised to redefine how project managers forecast and plan. Predictive analytics use historical data and AI algorithms to predict future outcomes in project management. These tools can analyze past project data to forecast potential delays or cost overruns, allowing project managers to proactively make decisions to steer the project back on course before issues become critical.

However, it's essential to consider the potential ethical implications of using AI, such as data privacy or algorithmic bias. The integration of AI provides an additional layer of sophistication by learning from data patterns to improve predictions' accuracy continuously. As AI technology advances, these tools will become more adept at handling complex data sets, providing insights that are not only reactive but also proactive, offering recommendations for action that are data-driven and tailored to the unique circumstances of each project.

As project documentation and reporting continue to evolve, integrating these technologies into everyday project management practices is becoming indispensable. By staying ahead of these trends, you, as a project manager, can leverage these innovations to drive efficiency, enhance stakeholder engagement, and deliver successful projects with confidence and precision. These advancements represent the beginning of what is possible in project management as technology expands the boundaries of what can be achieved.

8.6 Preparing for the Impact of Regulatory Changes on Project Management

In the intricate tapestry of project management, staying ahead of regulatory changes is akin to a ship's captain navigating through foggy waters. Particularly in healthcare, finance, and construction industries, regulations can change with little warning, significantly impacting project management practices. For instance, changes in data protection laws in finance have required projects to implement stricter security measures, leading to increased project costs and timelines. Safety protocols in construction or patient privacy in healthcare have also seen significant changes, affecting project planning and execution. Each regulation carries the weight of compliance, with hefty penalties for non-compliance. Thus, understanding these regulations and preparing for their impact is a legal and strategic imperative.

Anticipating regulatory changes requires a proactive approach. Staying informed through industry news, regulatory announcements, and professional networks is crucial. Engaging with industry associations can also provide insights into upcoming regulatory trends. Additionally, maintaining a dialogue with legal and compliance experts within your organization can give early warnings of changes that could impact your projects. For project managers, this foresight allows for swift adaptation of project scopes and objectives, ensuring that projects remain compliant while still achieving their goals. Consider, for example, introducing new environmental regulations that affect a construction project. By anticipating these changes, you can adjust procurement strategies to comply with new sustainability standards, avoiding delays and potential legal issues. This proactive approach ensures compliance and instills a sense of security and preparedness in your project management roles.

Strategies for ensuring compliance with new regulations include comprehensive training programs for project teams. These programs should cover the specific regulations relevant to the project and include updates whenever new rules are announced. Moreover, revising documentation practices to meet legal standards is critical. This might involve updating templates for project reporting or contracts to ensure all necessary compliance information is accurately captured and clearly stated. For instance, if a new regulation requires projects to report on their environmental impact, you might need to add a section to your project reports explicitly addressing this. Implementing compliance checks throughout the project lifecycle is another effective strategy. Regular audits and reviews can help catch potential compliance issues early, allowing for timely corrections before they escalate into more significant problems.

The impact of international regulations on global projects adds another layer of complexity. For project managers handling international projects, understanding the regulatory environment in each country is essential. This may involve working with local legal teams to navigate the specific legal requirements and cultural nuances of business in different regions. For instance, data protection laws vary significantly between the European Union and the United States and failing to understand these can lead to compliance breaches. Effective management of these variations requires a flexible approach to project planning and a deep understanding of international law as it applies to project activities. It's also important to consider the potential challenges, such as language barriers or differences in business practices when managing projects across borders.

Case studies on regulatory adaptation offer valuable lessons on how companies have successfully navigated regulatory changes. One notable example involves a multinational pharmaceutical company that faced significant changes in clinical trial regulations in multiple countries. By establishing a dedicated regulatory compliance team within the project management office, the company could stay abreast of changes and adapt its project plans accordingly. This proactive approach ensured compliance and minimized disruptions to the clinical trials, allowing the company to bring new drugs to market without significant delays. Another example comes from a financial services firm that had to adjust its data management practices in response to new data protection laws. By implementing robust data encryption and audit trails, the firm complied with the latest regulations and enhanced its reputation for data security, strengthening customer trust.

These examples underscore the importance of agility and informed decision-making in project management. As regulations evolve, your ability to anticipate and adapt to these changes will be crucial in guiding your projects to successful and compliant completions.

In wrapping up this exploration of regulatory impacts on project management, we've delved into the necessity of anticipating changes, strategies for ensuring compliance, and the global complexities of international regulations. Each aspect plays a pivotal role in molding project management practices that are not only compliant but also robust and responsive to the dynamic regulatory landscape. As we turn the page to the next chapter, the focus will shift to another critical facet of modern project management, further building on the foundation we've established here. This ongoing journey into the depths of project management is designed to equip you with the knowledge and skills needed to navigate the complexities of this ever-evolving field.

The Project Management Blueprint Review

Title: Keeping the Project Management Game Alive

Now you have everything you need to become a great project manager; it's time to share what you've learned and help other new project managers find the same guidance.

By sharing your honest thoughts about this book on Amazon, you'll show other beginners where they can find the help they're seeking. You'll pass on project management excitement to others just starting.

Thank you for being so helpful. Project management grows stronger when we share what we know – and you're helping us do just that.

>>> Click here to leave your review on Amazon.<<<

Or scan the QR Code to help us out!

Remember, your review doesn't have to be long or fancy. Just a few words about what you liked or found helpful can make a big difference to someone trying to learn about project management.

Maybe you found the helpful process, or perhaps the step-by-step guides helped you understand complex ideas. Whatever helped you, sharing it could be exactly what another reader needs to hear.

By leaving a review, you're helping other readers and supporting the project management community. You're showing that there are helpful resources out there for people who want to learn.

So take a minute to share your thoughts. Your review could be why someone else decides to start their project management journey. And who knows? The person you help today might be the one leading an amazing project tomorrow!

Thank you for being part of our project management family. Together, we're making the world of projects run a little smoother, one manager at a time. **Next up is the conclusion and Top Definitions used in Project Management!**

Chapter 9

Conclusion

I wanted to create a guide I wish I had when I first started, and I truly hope this book benefited you in learning multiple concepts. From the foundational concepts in the early chapters to the advanced strategies and tools discussed later, each section was carefully designed to build upon the previous, ensuring that you, the reader, gain a comprehensive understanding of excelling in project management.

We've covered essential project planning, examined the details of methodologies like Agile and Waterfall, and explored the potential of cutting-edge technologies. It's important to remember that the role of technology in project management cannot be overstated—it is the force that drives us forward, enhancing efficiency and opening new avenues for innovation. Embracing tools from simple project management software to complex AI and machine learning applications is critical in today's fast-evolving landscape.

The significance of certification in your career cannot be understated. As we discussed, obtaining certifications such as PMP or Agile is not just about adding a credential to your resume—it's about deepening your knowledge, expanding your professional network, and enhancing your strategic thinking capabilities. These certifications pave the way for career advancement and open doors to new opportunities.

I encourage you to take proactive steps toward advancing your project management career. Whether that means pursuing certification, implementing the strategies we've discussed in your current projects, or dedicating yourself to continuous learning and professional networking, each step you take is a building block in your career path.

Looking ahead, the future of project management is bright and boundless. Envision yourself as a leader who manages projects, drives innovation, champions sustainability, and bridges cultural divides in a globalized market. The skills you've acquired here are your toolkit for success across any sector—technology, construction, healthcare, or finance.

I also encourage you to participate actively in the project management community. Join forums, attend conferences, and engage in online communities. These platforms are invaluable for sharing experiences, gaining insights, and contributing to the broader conversation on project management.

Reflecting on my journey in project management, I am reminded of the challenges and immense fulfillment this career path offers. I hope 'The Project Management Blueprint' serves as a guide and a companion in your ongoing journey. May the lessons and insights you've gathered here inspire you to reach new heights in your career and, ultimately, contribute to shaping the future of project management.

Thank you for joining me on this journey. I wish you all the success you need as you continue to grow, innovate, and lead in the ever-evolving world of project management. I can't wait to see you on the field!

Chapter 10

Top 100 Definitions Used In Project Management

This is a list of the top 100 definitions you'll use throughout your career in Project Management

Agile

Agile is an iterative approach to project management and software development that emphasizes flexibility, collaboration, and rapid delivery. It breaks projects into smaller, manageable sprint units, allowing for frequent reassessment and adaptation.

Agile Release Train (ART)

An Agile Release Train is a long-lived team of 50-125 individuals, typically developing and delivering solutions incrementally using a cadence of Program Increments (PIs).

Analogous Estimating

Analogous Estimation is a technique for estimating the duration or cost of an activity or project using historical data from a similar activity or project. It's beneficial when detailed information about the current project is limited.

Assumption

An assumption in the planning process is considered valid, accurate, or specific without empirical proof or demonstration. Identifying and documenting assumptions is essential for project planning and risk assessment.

Backlog

A Backlog is a prioritized list of features, user stories, tasks, or requirements for a project. In Agile methodologies, it's often referred to as the Product Backlog.

Baseline

A Baseline is the approved version of a project plan component, used as a basis for comparison to track progress. Standard baselines include scope, schedule, and cost.

Bottom-up Estimating

Bottom-up Estimating is a method of estimating project duration or cost by aggregating the estimates of the lower-level components of the Work Breakdown Structure.

Budget at Completion (BAC)

The budget at Completion is the total estimated cost of the project when finished. It's a key component of Earned Value Management.

Burndown Chart

A Burndown Chart is a graphical representation of work left to do versus time. It helps teams track progress and predict the likelihood of achieving the sprint goal.

Change Control

Change Control manages changes to the project scope, schedule, or budget. It involves documenting, evaluating, and approving changes before implementation.

Change Request

A Change Request is a formal proposal to modify any deliverable, project document, or baseline.

Closing Process Group

The Closing Process Group consists of those processes performed to formally complete or close a project, phase, or contract.

Constraint

Constraint is a limiting factor that affects the execution of a project, program, portfolio, or process. Typical constraints include time, cost, and scope.

Contingency Reserve

A contingency Reserve is an amount in the project budget or schedule allocated for identified risks that are accepted and for which contingent or mitigation responses are developed.

Cost-Benefit Analysis

Cost-benefit analysis is a systematic approach to estimating the strengths and weaknesses of alternatives to determine the best approach to achieving benefits while preserving savings.

Critical Chain Project Management (CCPM)

Critical Chain Project Management is a method of planning and managing projects emphasizing the resources required to execute project tasks. It differs from the Critical Path method in that duration buffers are added to the project schedule.

Critical Path

The critical path is the most extended sequence of dependent tasks in a project, determining the shortest time to complete the project.

Critical Success Factors (CSF)

Critical Success Factors are the key areas of activity in which favorable results are necessary for a manager to reach his/her goals. In project management, CSFs are essential factors for project success.

Deliverable

A Deliverable is any unique and verifiable product, result, or capability produced to complete a process, phase, or project. Deliverables can be tangible or intangible.

Deliverable Acceptance

Deliverable Acceptance is receiving formal approval from the client or stakeholder that a deliverable meets the agreed-upon criteria.

Dependency

A Dependency in project management refers to the relationship between two activities where the start or completion of one activity depends on the start or completion of another.

Dependency Types (FS, SS, FF, SF)

Dependency Types describe the relationships between tasks. These include Finish-to-Start (FS), Start-to-Start (SS), Finish-to-Finish (FF), and Start-to-Finish (SF).

Earned Value Management (EVM)

Earned Value Management is a technique for objectively measuring project performance and progress. It integrates scope, schedule, and cost data.

Estimation Techniques

Estimation Techniques are methods to forecast the time, cost, and resources required for a project or task. These can include analogous, parametric, three-point, and bottom-up estimating.

Executing Process Group

The Executing Process Group consists of those processes performed to complete the work defined in the project management plan and satisfy the project specifications.

Float (or Slack)

Float, also known as slack, is the amount of time a task can be delayed without causing a delay to subsequent tasks or the project completion date.

Gantt Chart

A Gantt chart visually represents a project schedule, displaying tasks as horizontal bars on a timeline.

Initiating Process Group

The Initiating Process Group consists of those processes performed to define a new project or a new phase of an existing project by obtaining authorization to start the project or phase.

Iteration

An Iteration is a time-boxed period during which specific work is completed and made ready for review. It's a fundamental concept in Agile methodologies, similar to a Sprint in Scrum.

Kanban

Kanban is a visual method for managing work as it moves through a process, typically using cards on a board to represent work items.

Kickback

A Kickback in project management refers to the rejection or return of a deliverable or work item due to quality issues or non-conformance with requirements.

Kickoff Meeting

A Kickoff Meeting is the first meeting with the project team and stakeholders at the beginning of a project. It sets the tone for the entire project and aligns everyone on project goals, roles, and expectations.

Lean Project Management

Lean Project Management is an approach that focuses on delivering value to the customer while minimizing waste. It emphasizes efficiency, continuous improvement, and respect for people.

Lessons Learned

Lessons learned are the knowledge gained during a project that shows how project events were addressed or should be addressed in the future.

Lessons Learned Register

A Lessons Learned Register is a project document used to record knowledge gained during a project, which can be used to improve future performance.

Milestone

A milestone is a significant point or event in a project timeline, often marking the completion of a primary deliverable or phase.

Monitoring and Controlling Process Group

The Monitoring and Controlling Process Group consists of those processes required to track, review, and regulate the progress and performance of the project, identify any areas in which changes to the plan are needed, and initiate the corresponding changes.

Monte Carlo Simulation

Monte Carlo Simulation is a computerized mathematical technique that allows people to account for risk in quantitative analysis and decision-making.

Organizational Breakdown Structure (OBS)

An Organizational Breakdown Structure is a hierarchical representation of the project organization that illustrates the relationship between project activities and the organizational units that will perform those activities.

Parametric Estimating

Parametric Estimating is an estimating technique that uses a statistical relationship between historical data and other variables to calculate an estimate for activity parameters, such as cost, budget, and duration.

PERT (Program Evaluation and Review Technique)

PERT is a method to analyze the tasks involved in completing a given project, especially the time needed to complete each task and identify the minimum time required to complete the total project.

Planning Process Group

The Planning Process Group consists of those processes performed to establish the total scope of the effort, define and refine the objectives, and develop the course of action required to attain those objectives.

PMBOK (Project Management Body of Knowledge)

PMBOK is a standard terminology and guidelines for project management published by the Project Management Institute (PMI).

Portfolio Management

Portfolio Management is the centralized management of one or more portfolios to achieve strategic objectives. It includes identifying, prioritizing, authorizing, managing, and controlling projects, programs, and other related work.

Precedence Diagramming Method (PDM)

The Precedence Diagramming Method is a technique used in project management to construct a schedule network diagram that uses boxes or rectangles, referred to as nodes, to represent activities and connect them with arrows that show the dependencies.

Program Management

Program Management is managing several related projects, often to improve an organization's performance.

Project

A Project is a temporary endeavor to create a unique product, service, or result. It has a defined beginning and end in time and, therefore, defined scope and resources.

Project Charter

A Project Charter is a formal, typically short document that outlines the project's objectives, scope, stakeholders, and high-level requirements. It authorizes the project and gives the project manager the authority to apply organizational resources to project activities.

Project Closure

Project Closure is finalizing all activities across all project management process groups to complete the project formally.

Project Communication Plan

A Project Communication Plan defines the communication requirements for the project and how information will be distributed.

Project Lifecycle

The Project Lifecycle is the series of phases a project goes through, from initiation to closure. Typically, it includes initiation, planning, execution, monitoring and controlling, and closing phases.

Project Management

Project Management is the application of knowledge, skills, tools, and techniques to project activities to meet the project requirements.

Project Management Information System (PMIS)

A Project Management Information System is a software tool used to support various aspects of project management, such as planning, scheduling, resource allocation, and control.

Project Management Office (PMO)

A Project Management Office is a department or group that defines and maintains project management standards.

Project Management Plan

The Project Management Plan is a formal, approved document that defines how the project is executed, monitored, and controlled.

Project Management Triangle

The Project Management Triangle, also known as the Triple Constraint, illustrates the relationship between three primary forces in a project: scope, time, and cost.

Project Manager

A project manager is the person the performing organization assigns to lead the team and is responsible for achieving the project objectives.

Project Network Diagram

A Project Network Diagram is a schematic display of the logical relationships among project activities.

Project Portfolio Management (PPM)

Project Portfolio Management is the centralized management of processes, methods, and technologies that project managers and offices use to analyze and collectively manage current or proposed projects.

Project Schedule

A Project Schedule is a document that lists a project's tasks, milestones, and deliverables, along with their planned start and finish dates.

Project Scope

Project Scope is the work that needs to be accomplished to deliver a product, service, or result with the specified features and functions.

Project Sponsor

A Project Sponsor is a senior executive in an organization responsible for providing resources and support for the project. They champion the project at the highest level and are ultimately accountable for its success.

Project Team

The Project Team consists of individuals who work together to achieve project objectives.

Quality Assurance (QA)

Quality Assurance is auditing the specific project requirements and the results from quality control measurements to ensure appropriate quality standards and operational definitions are used.

Quality Control (QC)

Quality Control is the process of monitoring and recording results of executing quality activities to assess performance and recommend necessary changes.

Quality Management

Quality Management in project management involves the processes required to ensure that the project will satisfy the needs for which it was undertaken. It includes quality planning, quality assurance, and quality control.

RACI Matrix

RACI (Responsible, Accountable, Consulted, Informed) Matrix is a responsibility assignment chart that maps out every task, milestone, or decision and assigns roles to project team members.

Resource Allocation

Resource allocation involves assigning and managing the resources (people, equipment, materials) needed to complete project tasks.

Resource Histogram

A Resource Histogram is a bar chart showing the time a resource is scheduled to work over a specific period.

Resource Leveling

Resource Leveling is a technique in project planning and scheduling that aims to optimize the allocation of resources and resolve resource conflicts or over-allocations.

Risk

A Risk is an uncertain event or condition that, if it occurs, has a positive or negative effect on one or more project objectives.

Risk Management

Risk management identifies, assesses, and controls potential issues that could affect a project.

Risk Register

A Risk Register is a document used to identify, analyze, and track potential risks in a project.

Risk Response Strategies

Risk Response Strategies are approaches to identifying risks, including avoidance, transference, mitigation, and acceptance.

Rolling Wave Planning

Rolling Wave Planning is an iterative planning technique in which the work to be accomplished in the near term is planned in detail, while the work in the future is scheduled for a higher level.

Scope

Scope defines the boundaries of a project, outlining what is and isn't included in the project's deliverables.

Scope Baseline

The Scope Baseline is the approved version of a scope statement, Work Breakdown Structure (WBS), and its associated WBS dictionary.

Scope Creep

Scope Creep refers to the uncontrolled expansion of project scope without adjustments to time, cost, and resources.

Scope Statement

A Scope Statement is a document that provides a detailed description of the project and product, service, or result.

Scrum

Scrum is an Agile framework for project management that emphasizes teamwork, accountability, and iterative progress.

Sprint

A Sprint is a fixed time box within which specific work must be completed and made ready for review. Typically lasting 1-4 weeks, Sprints are fundamental to Agile methodologies, particularly Scrum.

Stakeholder

Stakeholders are individuals or groups interested in or affected by the project's outcome.

Stakeholder Analysis

Stakeholder Analysis is the process of identifying and analyzing stakeholders, and systematically gathering and analyzing qualitative information to determine whose interests should be considered throughout the project.

Stakeholder Engagement

Stakeholder Engagement refers to interacting with, communicating with, and influencing project stakeholders to support the project's interests.

Stakeholder Register

A Stakeholder Register is a document containing project stakeholders' identification, assessment, and classification.

Statement of Work (SOW)

A Statement of Work is a narrative description of products or services to be supplied by the project.

Task Duration

Task Duration is the time required to complete a specific task or activity.

Three-Point Estimate

A Three-Point Estimate is a technique used to estimate cost or duration by applying an average of optimistic, pessimistic, and most likely scenarios.

Time-Boxed

Time-Boxed refers to a project or activity with a fixed duration, regardless of whether the planned scope has been achieved.

Use Case

A Use Case is a list of actions or event steps typically defining the interactions between a role (actor) and a system to achieve a goal.

Velocity

In Agile project management, Velocity measures the work a team can tackle during a single sprint.

Waterfall Model

The Waterfall Model is a linear sequential approach to project management where progress flows steadily downwards through phases like conception, initiation, analysis, design, construction, testing, and maintenance.

What-If Scenario Analysis

What-if scenario Analysis is the process of evaluating scenarios to predict their effect on project objectives.

Work Breakdown Structure (WBS)

A Work Breakdown Structure is a hierarchical decomposition of the total scope of work to be carried out by the project team.

Work Package

A Work Package is a deliverable or project work component at the lowest level of the Work Breakdown Structure.

Workload

Workload refers to the work assigned to or expected from a team or individual over a given period.

Workaround

A Workaround is a response to an adverse risk that has occurred. It's distinguished from a contingency plan in that a workaround is not planned for the risk occurrence.

Workflow

Workflow is the sequence of processes through which a piece of work passes from initiation to completion.

Workflow Diagram

A Workflow Diagram is a graphical representation of a process or workflow, typically using standardized symbols to represent different actions or steps. It visually depicts a process's sequence of operations, interactions, and decision points. Project Managers use Workflow Diagrams to analyze, design, and communicate complex processes, identify inefficiencies, and improve overall project workflow.

WBS Dictionary

The WBS Dictionary is a document that provides detailed information about each component in the Work Breakdown Structure (WBS). It typically includes a description of the work, responsible party, schedule, resources required, cost estimates, quality requirements, and technical references. The WBS Dictionary is an essential tool for Project Managers to ensure a clear understanding of project scope elements and to support accurate planning and execution of project tasks.

Chapter 11

References

1. Green Frog Computing. (n.d.). The importance of IT project management in modern business. https://www.greenfrogcomputing.co.uk/the-importance-of-it-project-management/

2. Float. (n.d.). Agile vs. waterfall: 10 key differences between the two. https://www.float.com/resources/agile-vs-waterfall#:~:text=Delivery%3A%20Agile%20allows%20for%20quick,changes%20during%20the%20development%20process

3. The Digital Project Manager. (2024). 15 best project management software picked for 2024. https://thedigitalprojectmanager.com/tools/best-project-management-software/

4. Project Management Institute. (n.d.). Project management case studies. https://www.pmi.org/business-solutions/case-studies

5. LeanDog. (n.d.). Agile implementation: The approach, process and models. https://www.leandog.com/agile-process-models-guide

6. The Digital Project Manager. (2024). How to develop a project plan: Template & example [2024]. https://thedigitalprojectmanager.com/projects/managing-schedules/project-plan-guide/

7. UpGrad. (n.d.). Top 20 project management case studies [with examples]. https://www.upgrad.com/blog/project-management-case-studies-with-examples/

8. Coursera. (n.d.). How to manage project risk: A 5-step guide. https://www.coursera.org/articles/how-to-manage-project-risk

9. Forbes. (2024). 7 essential project management techniques in 2024. https://www.forbes.com/advisor/business/project-management-techniques/

10. SAP. (n.d.). What is ERP? The essential guide. https://www.sap.com/products/erp/what-is-erp.html

11. Six Seconds. (2014, January 14). Emotional intelligence improves leadership at FedEx. https://www.6seconds.org/2014/01/14/case-study-emotional-intelligence-people-first-leadership-fedex-express/

12. ProjectManager. (n.d.). Using earned value management to measure project performance. https://www.projectmanager.com/blog/using-earned-value-management-to-measure-project-performance

13. The Digital Project Manager. (2024). 15 best project management software picked for 2024. https://thedigitalprojectmanager.com/tools/best-project-management-software/

14. Harvard Business Review. (2023, February). How AI will transform project management. https://hbr.org/2023/02/how-ai-will-transform-project-management

15. The Digital Project Manager. (n.d.). 10 best remote project management tools reviewed for ... https://thedigitalprojectmanager.com/tools/remote-project-management-tools/

16. Institute of Chartered Accountants in England and Wales. (n.d.). Blockchain case studies | Technology. https://www.icaew.com/technical/technology/blockchain-and-cryptoassets/blockchain-articles/blockchain-case-studies

17. Project Management Institute. (n.d.). The five stages of team development and the role ... https://www.projectmanagement.com/blog-post/15192/the-five-stages-of-team-development-and-the-role-of-the-project-manager

18. Wellable. (n.d.). 7 effective conflict resolution techniques in the workplace. https://www.wellable.co/blog/conflict-resolution-techniques-in-the-workplace/

19. PM Exam SmartNotes. (n.d.). 5 motivational theories that a project manager must know ... https://pmexamsmartnotes.com/motivational-theories-for-project-manager/

20. Kahootz. (n.d.). Stakeholder engagement strategy guide with examples. https://www.kahootz.com/how-to-create-an-effective-stakeholder-engagement-strategy/

21. PMO365. (n.d.). Sustainability in project management: A complete guide. https://pmo365.com/blog/sustainability-in-project-management-a-complete-guide

22. British University in Dubai. (n.d.). Exploring the impact of globalization on project management. https://bspace.buid.ac.ae/items/ff403704-2a8e-41fc-8020-4efd1fefc7db

23. Productive. (n.d.). Project financial management: Key strategies for success. https://productive.io/blog/project-financial-management/

24. Project Management Academy. (n.d.). CAPM vs PMP certification: Your comprehensive guide. https://projectmanagementacademy.net/resources/blog/pmp-vs-capm/

25. Forbes. (n.d.). PMP exam prep resources to help you get certified. https://www.forbes.com/advisor/education/certifications/pmp-exam-prep/

26. KnowledgeHut. (n.d.). Top 10 benefits of PMI membership. https://www.knowledgehut.com/blog/project-management/benefits-of-pmi-membership

27. Planview. (n.d.). A guide to networking and project management. https://blog.planview.com/a-guide-to-networking-and-project-management/

28. The Digital Project Manager. (2023). 5 emerging project management trends of 2023. https://thedigitalprojectmanager.com/industry/reports/project-management-trends/

29. Harvard Business Review. (2023, February). How AI will transform project management. https://hbr.org/2023/02/how-ai-will-transform-project-management

30. Project Management Institute. (n.d.). Managing cross cultural differences in projects. https://www.pmi.org/learning/library/managing-cross-cultural-differences-projects-6736

31. Engineers Edge. (n.d.). Project management basics online training - 4 P D H . https://www.engineersedge.com/engineering/Services_Directory/Professional_Training/project_management_basics_online_training__4pdh_10140.htm

32. Tasko Consulting. (n.d.). What is agile software development. https://taskoconsulting.com/what-is-agile-software-development/

33. McDargh, E. (n.d.). What is transformational leadership? https://www.eileenmcdargh.com/post/what-is-transformational-leadership

34. Cwallet. (n.d.). Using blockchain to combat disinformation in the digital era. https://blog.cwallet.com/using-blockchain-to-combat-disinformation-in-the-digital-era/

35. Bodemer, O. (2023). Decentralized innovation: Exploring the impact of blockchain technology in software development. https://doi.org/10.36227/techrxiv.24456100

36. Project Management Institute. (2021). A guide to the project management body of knowledge (PMBOK guide) (7th ed.). Project Management Institute.

37. AXELOS. (2017). Managing successful projects with PRINCE2 (6th ed.). The Stationery Office.

38. Agile Alliance. (n.d.). Agile 101. https://www.agilealliance.org/agile101/

39. Kerzner, H. (2017). Project management: A systems approach to planning, scheduling, and controlling (12th ed.). John Wiley & Sons.

40. Mulcahy, R. (2018). PMP exam prep: Accelerated learning to pass the Project Management Professional (PMP) exam (9th ed.). RMC Publications.

41. Berkun, S. (2008). Making things happen: Mastering project management. O'Reilly Media.

Made in the USA
Coppell, TX
23 October 2024